MRCPsych PARTS I AND II:
INDIVIDUAL STATEMENTS
AND EMI PRACTICE EXAMS

MRCPsych PARTS I AND II:
INDIVIDUAL STATEMENTS AND EMI PRACTICE EXAMS

Justin Sauer MBBS BSc MRCPsych
Specialist Registrar, Maudsley Hospital, London, UK

Hodder Arnold

A MEMBER OF THE HODDER HEADLINE GROUP

First published in Great Britain in 2005 by
Hodder Education, a member of the Hodder Headline Group,
338 Euston Road, London NW1 3BH

http://www.hoddereducation.co.uk

Distributed in the United States of America by
Oxford University Press Inc.,
198 Madison Avenue, New York, NY 10016
Oxford is a registered trademark of Oxford University Press

Whilst the advice and information in this book are believed to be true and
accurate at the date of going to press, neither the author[s] nor the publisher
can accept any legal responsibility or liability for any errors or omissions
that may be made. In particular (but without limiting the generality of the
preceding disclaimer), every effort has been made to check drug dosages;
however, it is still possible that errors have been missed. Furthermore,
dosage schedules are constantly being revised and new side-effects
recognized. For these reasons the reader is strongly urged to consult the
drug companies' printed instructions before administering any of the drugs
recommended in this book.

British Library Cataloguing in Publication Data
A catalogue record for this book is available from the British Library

Library of Congress Cataloging-in-Publication Data
A catalog record for this book is available from the Library of Congress

ISBN-10:   0 340 90471 2
ISBN-13:   978 0 340 90471 8

1 2 3 4 5 6 7 8 9 10

Commissioning Editor: Clare Christian
Project Editor: Clare Patterson
Production Controller: Jane Lawrence
Cover Design: Nichola Smith
Indexer: Laurence Errington

Typeset in 9.5 on 12pt Rotis Serif by Phoenix Photosetting, Chatham, Kent.
Printed and bound in Great Britain by CPI Bath.

What do you think about this book? Or any other Hodder Arnold title?
Please visit our website at www.hoddereducation.co.uk

Dedicated to Nina, Jay and Kimby

CONTENTS

FOREWORD

The MRCPsych is a very fair examination. I know that there will be times during your personal preparation when you will view the exam as an enormous and impossibly difficult hurdle. 'What kind of twisted sadist,' you will ask, 'would dream up such impossible questions whose answers cannot be found even in the biggest textbooks?' As comfort, I would ask you to consider my (albeit second-hand) experience of the exam, gained during 9 years as Tutor for a scheme with 83 trainees. As cohort after cohort of SHOs went through both parts of the exam, my constant observation was that the seriousness with which a candidate approached their preparation for the exam was absolutely linked to their chances of success. Indeed, when I saw brilliant candidates fail, the reason was always clear – they simply hadn't prepared sufficiently. Far less naturally talented folk sailed through every time if they put the effort in during the revision period. So, if 'preparation, preparation, preparation' is the secret to success in the MRCPsych, how should you go about it? The most time-efficient way to cover the theoretical material needed for the exam is to work through questions like the ones in this book. You will quickly learn which are the areas where your knowledge base is weakest and can direct your textbook reading appropriately. You will also be able to judge how far your revision still needs to go. If you can, study the questions with a colleague or a group of trainees who are preparing for the exam. You will find that the discussions that result from attempts to answer some of the questions will throw up areas of knowledge that are completely new to you. Further, you are much more likely to remember the answers to these questions in a few weeks' time if they come up in the real exam situation if you have had a discussion with friends about what the right answer really is. Remember that there are only so many questions that can be set in the area and that a tremendous amount of recycling of questions happens. Hence, there's a very good chance that some of the questions in this book will appear in the exam paper at the next sitting. So – enjoy the book and use it as part of your preparation, happy in the knowledge that if you've prepared yourself adequately you can expect to pass.

Robert Howard, Institute of Psychiatry, London

ACKNOWLEDGEMENTS

I would like to thank my friends and colleagues who have encouraged me over the years, especially Robert Howard, Dinesh Bhugra, Oyedeji Ayonrinde and Naji Tabet.

I am grateful also to my parents for giving me opportunities and for their support.

Thank you to Hodder and in particular to Clare Christian for her enthusiasm.

HOW TO USE THIS BOOK

For many, the thought of exams is distressing. Right now they probably feel like expensive, stressful and unnecessary obstacles hindering your rightful career progression.

This is an 'exam adjustment reaction'! It is a normal response and will pass. The good news is that almost everyone is in the same position. Attempting to balance busy clinical jobs with family and domestic commitments is an achievement in itself. Fitting examination practice into the equation can be challenging.

Whilst reading textbooks is an important component to learning any subject, practising questions is a vital part of passing examinations. With the above in mind, question practice should be snuck into little moments throughout the busy day. As I was advised to practise as many questions as I could lay my hands on, I strongly advise you do the same. Those questions that have you flummoxed should prompt you to look in your textbooks to find out more. This will help you remember for the next time.

Included in this book are six practice papers comprising over 1000 questions. Section 1 contains the Part I papers and Section 2 the Part II papers. I would recommend sitting them as mock (timed) examinations if possible, but they can be used as revision aids or to stimulate further reading. Most questions are accompanied by written explanations. I have included references where appropriate should you feel the need to look at papers in more detail. Part II candidates are expected to have retained Part I knowledge, so will benefit from practising all six papers. Clinical and basic science questions are combined in the Part II papers to prevent hypersomnia. Part I candidates will find it useful to peruse the Part II papers, and will be able to use the book again for Part II preparation.

Revision advice

Make no excuses for delaying the start of your revision. Be sceptical about tales of those who did not revise for their exams and passed easily, or those who only dedicated 6 weeks to revision. You know yourself and how long you will need. Get into a routine of revising regularly and of doing as many practice exam questions as you can. Practising questions allows for self-assessment and recognition of areas that need further work. For many, question-based revision is more stimulating then ploughing through textbooks. Also, by practising questions repeatedly, you get into the examination mode and the examiners' mind set, so that you begin to see and predict questions when you are reading texts or journals.

Remember the more you practise, the luckier you will be in the actual exam. I hope you find these papers helpful and stimulating in your preparation for your forthcoming examinations.

Currently the Part I written paper is a 90-minute examination comprising 133 individual statement items (ISIs) and 10 themed extended matching items (EMIs). The Royal College of Psychiatrists recommends spending 60 minutes on the ISIs and 30 minutes on the EMIs.

The Part II papers consist of 165 ISIs and 5 themed EMIs. In the 90-minute examination it is recommended that you spend 75 minutes on the ISIs and 15 minutes on the EMIs. I have added an extra EMI to each Part II paper to give you more practice.

Good luck.

J.S.

SECTION 1:
PART I MRCPsych

PAPER 1

INDIVIDUAL STATEMENTS

1   D4 receptor blockade in schizophrenia is associated with an antipsychotic effect.   T | F

2   Amisulpride is a selective D1 receptor antagonist.   T | F

3   Bupropion is a nicotinic receptor agonist.   T | F

4   Bupropion is contraindicated in patients with a history of eating disorders.   T | F

5   Clozapine is effective in 80 per cent of patients with schizophrenia who do not respond to conventional neuroleptic therapy.   T | F

6   Lithium has been shown to affect neurogenesis and to increase total grey matter content.   T | F

7   Perfectionism as a personality trait is rarely associated with anorexia nervosa.   T | F

8   Secondary memory has a limited capacity.   T | F

9   The presence of comorbid anxiety in depression is a predictor of suicidality.   T | F

10   Psychotic symptoms are present in 5–10 per cent of patients with severe depression.   T | F

11   Suicide rates increase during wartime.   T | F

12   In psychotherapy, 'clarification' offers new formulations of unconscious meaning and motivation.   T | F

13   Psychomotor retardation has been identified as a predictor of a good response to ECT.   T | F

14   Hypoxia increases the permeability of the blood–brain barrier.   T | F

15   A reduction in TV, video and video game usage in 9-year-old children reduces aggressive behaviour.   T | F

16   Approximately 30 per cent of young women with a serious drinking problem have a significant eating disorder at some time.   T | F

17   Ecstasy (MDMA) is a dissociative anaesthetic.   T | F

18   MDMA (Ecstasy) damages the serotonergic nerve terminal.   T | F

19   Tetrachloroethylene is most commonly abused by middle-aged men.   T | F

20   Paracetamol overdose accounts for 27 per cent of suicides by drug ingestion.   T | F

21   Susto is a condition in which it is believed a sudden fright can make the soul leave the body.   T | F

22   BITE is a self-report questionnaire used in bulimia nervosa.   T | F

23   Depression is more commonly associated with bulimia than with anorexia.   T | F

24   Disulfiram exacerbates schizophrenia.   T | F

25   Double depression describes episodes of major depression superimposed on cyclothymia.   T | F

26   Episodes of depression are usually shorter than episodes of mania.   T | F

27   The term 'atypical neuroleptics' relates to the failure to produce catalepsy in laboratory animals.   T | F

28   Rabbit syndrome is a movement disorder affecting the lips and peri-oral muscles.   T | F

29   Akinesia is associated with neuroleptic usage.   T | F

30   The Abnormal Involuntary Movement Scale (AIMS) scale is a measure of mental-health-related social disability.   T | F

31   Flumazenil is a benzodiazepine receptor antagonist.   T | F

32   Carbamazepine increases the levels of sodium valproate.   T | F

33   Caffeine causes global cerebral vasoconstriction.   T | F

34   Flashbacks of hallucinations following hallucinogenic usage can occur for months afterwards.   T | F

35   Veraguth's fold is a feature of schizophrenia.   T | F

36   Mitral valve prolapse is associated with panic disorders.   T | F

37   People with affective disorders may be at an increased risk of developing dementia.   T | F

38   Tricyclic antidepressants enhance cardiac conduction.   T | F

39   Khat usage is associated with an amphetamine-like psychosis.   T | F

40   Somatopagnosia is the inability of an individual to recognise a   T | F
      neurological deficit in themselves.

41   Memantine may help protect neurones from elevated glutamate   T | F
      levels.

42   Eicosapentanoic acid should be used with caution in   T | F
      schizophrenia.

43   Sixty per cent of patients with schizophrenia regularly smoke.   T | F

44   Abreaction can be performed with a slow infusion of diazepam.   T | F

45   Sildenafil is contraindicated in patients who have hereditary   T | F
      degenerative retinal disorders.

46   ICD-10 includes body dysmorphic disorder as part of   T | F
      hypochondriacal disorder.

47   Complications of trichotillomania include carpal tunnel syndrome.   T | F

48   Separation anxiety in children is greatest between the ages of 6   T | F
      and 9 months.

49   Tacrine can cause liver damage.   T | F

50   Atasia abasia is the reduced impulse to act and think.   T | F

51   Sodium valproate is well absorbed following an oral   T | F
      administration but is only 30 per cent protein bound.

52   Opiate antagonists are useful in self-injurious behaviour.   T | F

53   Obsessive–compulsive disorder has a prevalence of 2.5 per cent.   T | F

54   Successful selective serotonin re-uptake inhibitor treatment for   T | F
      obsessive–compulsive disorder means tricyclic antidepressants no
      longer have a role in this condition.

55   Zopiclone decreases slow-wave sleep.   T | F

56   Lithium reduces REM sleep.   T | F

57   Anaclitic object-choice occurs when the choice is based on the   T | F
      pattern of childhood.

58 Globus hystericus is an irrational fear of the world. T | F

59 Negative symptoms are a risk factor for tardive dyskinesia in patients receiving antipsychotic treatment. T | F

60 According to the World Health Organisation, 500 million people in the world are affected by depression at any one time. T | F

61 St Louis hysteria is the same as Briquet's syndrome. T | F

62 Living as a woman for a minimum of 1 year is the usual life test required for male-to-female transsexual gender reassignment. T | F

63 Advancing paternal age appears to be an independent risk factor for schizophrenia. T | F

64 Unilateral ECT is more potent than bilateral ECT and has more pronounced side effects. T | F

65 The EEG is useful in diagnosing Alzheimer's disease. T | F

66 Alcohol increases alpha activity on EEG tracings. T | F

67 Akathisia is associated with violence. T | F

68 Duodenal stenosis is a complication of anorexia nervosa. T | F

69 The 'stages of change' model includes the plateau phase. T | F

70 Schizophrenia is the most genetically determined psychiatric disorder. T | F

71 Dysthymia describes a persistent depressed state of at least 24 months' duration. T | F

72 Lithium passes poorly into breast milk. T | F

73 Post-partum recurrence of affective psychosis is estimated at 30–50 per cent. T | F

74 Renal patients on haemodialysis have a reduced suicide frequency. T | F

75 Body mass index is height (m^2)/weight (kg). T | F

76 Body fat distribution is waist (cm)/hip (cm). T | F

77 The psychiatrist should have a fiduciary relationship with his or her patients. T | F

78 Sialorrhoea occurs in neuroleptic malignant syndrome.   T | F

79 Patients with seasonal affective disorder can tend towards
    hypomania in the spring.   T | F

80 The coexistence of pseudodementia and dementia is uncommon.   T | F

81 The onset of grief 1 week after the event is abnormal.   T | F

82 Chronic fatigue syndrome is classified in ICD-10.   T | F

83 Oxazepam is a short-acting benzodiazepine (less than 6 hours).   T | F

84 Women make up 75 per cent of people diagnosed with borderline
    personality disorder.   T | F

85 'Borderline' in borderline personality disorder was coined to
    describe people on the borderline between psychosis and neurosis.   T | F

86 Counter-conditioning is classical conditioning that occurs through
    imagery and not real experiences.   T | F

87 The Premack principle is employed in behaviour therapy.   T | F

88 The Thematic Apperception Test (TAT) is a projective test
    consisting of a set of ten ink blots.   T | F

89 Speedballs are used by drug addicts.   T | F

90 LSD is a synthetic hallucinogen which can be injected
    intravenously.   T | F

91 Melatonin levels are high during the day and decrease during
    sleep.   T | F

92 Cognitive therapy helps the individual acquire coping strategies.   T | F

93 Suicide is a form of acting out.   T | F

94 Individuals who have attempted suicide once have a 27-fold
    greater risk of subsequently committing suicide than the general
    population.   T | F

95 Hyperprolactinaemia may increase the risk of venous
    thromboembolism in some patients taking antipsychotic
    medication.   T | F

96 Dialectical behaviour therapy was developed specifically to treat
    borderline personality disorder.   T | F

97　Anxiety disorders are the commonest psychiatric disorders seen in diabetic patients.　　T　F

98　Cognitive analytic therapy usually lasts for a minimum of 30 sessions.　　T　F

99　Deprivation is a cause of speech (language) delay.　　T　F

100　Clozapine causes agranulocytosis in 7 per cent of treated subjects.　　T　F

101　Narrow angle glaucoma is a contraindication for olanzapine.　　T　F

102　Jung described eight basic operations of the mind.　　T　F

103　Approximately two-thirds of those with anorexia nervosa have comorbid social phobia.　　T　F

104　Low serum phosphate is a complication of re-feeding in anorexia nervosa.　　T　F

105　T4 levels are usually increased in anorexia nervosa.　　T　F

106　Amitriptyline is a tertiary amine.　　T　F

107　Disulfiram reduces the plasma levels of tricyclic antidepressants.　　T　F

108　Liddle's classification of schizophrenia is based on factor analysis of cerebral blood flow patterns.　　T　F

109　Amisulpride is a pure D2 antagonist.　　T　F

110　Neuroleptic malignant syndrome has not been reported with clozapine or risperidone.　　T　F

111　In Kleinian theory, the depressive position is linked to manic defences.　　T　F

112　Follow-up studies of schizophrenia show a suicide rate of approximately 20 per cent.　　T　F

113　Conversion symptoms usually conform to a physician's idea of a disease.　　T　F

114　Gestalt psychology emphasises the importance of perceiving whole objects and proposes a number of principles to explain how we organise them.　　T　F

115　Progressive gait disorder is associated with normal pressure hydrocephalus.　　T　F

116   Features of Parson's sick role include wanting to remain unwell.   T F

117   Imperative hallucinations are of little clinical significance.   T F

118   First rank symptoms are absent in 20 per cent of patients with schizophrenia.   T F

119   According to studies of child temperament, 2 per cent of children are 'slow to warm up'.   T F

120   Memory is profoundly affected in Huntington's disease relative to other cognitive functions.   T F

121   Sodium lactate induces symptoms of panic.   T F

122   The suicide rate is higher amongst the never married than the widowed.   T F

123   The susceptibility of individuals to develop depression as a result of major stressful life events changes with age.   T F

124   Freud thought aggression was a fundamental human drive.   T F

125   ECT up-regulates 5HT2 receptors.   T F

126   Tetrabenazine is a known cause of depression.   T F

127   Anorexia nervosa has the highest mortality of any psychiatric disorder.   T F

128   The seizure threshold lowers during a course of ECT, particularly bilateral ECT.   T F

129   In ECT the seizure threshold is higher in women.   T F

130   The lifetime prevalence of a unipolar depressive episode is 17 per cent.   T F

131   The community prevalence of mania in the elderly is 1.8 per cent.   T F

132   ≥ 35 µg oestrogen daily might increase the risk of depression.   T F

133   Depression is associated with reduced REM latency.   T F

EMIs

1

A	Procyclidine	F	Fenfluramine
B	Clozapine	G	Benzodiazepines
C	Prednisolone	H	Chloral hydrate
D	Dantrolene	I	Disulfiram
E	Amphetamine	J	Lithium

A patient not known to you has taken a large overdose of their medication. Which is the most likely drug taken if the patient presents with the following symptoms and signs?

1   Irritability, elation, hyperactivity, dry mouth, reduced concentration, chest pain, arrhythmia, heart block, psychotic symptoms.

2   Slurred speech, poor co-ordination, somnolence, hypotension, hyporeflexia, confusion, coma, respiratory depression.

3   Delirium, drowsiness, tachycardia, arrhythmia, hypotension, hypersalivation, seizures.

4   Headache, rash, peripheral/optic neuropathy, mucous membrane injury, psychotic behaviour.

5   Diarrhoea, vomiting, confusion, drowsiness, tremor, vertical nystagmus, muscle rigidity.

2

A	Somatisation disorder	G	Munchausen's syndrome by
B	Conversion disorder		proxy
C	Hypochondriasis	H	Pseudopsychiatric syndrome
D	Body dysmorphic disorder	I	Malingering
E	Chronic pain syndrome	J	Factitious disorder
F	Munchausen's syndrome		

What is the most likely diagnosis in the following cases?

1   A 27-year-old woman with loss of sensation and power in her lower limbs. Neurological impairment does not map to specific dermatomes and she is not unduly concerned.

2   A 42-year-old bus driver, unemployed for 2 years, claims he is unable to walk following a fall on uneven pavement. Despite the absence of abnormalities on physical examination, he is suing the local authority for compensation.

3   A 21-year-old woman wants corrective surgery as she believes her eyes are spaced too widely apart.

3

A	Simple partial seizures	G	Isolated tonic or clonic seizures
B	Complex partial seizures		
C	Absence seizures (petit-mal)	H	Febrile convulsions
D	Tonic–clonic seizures	I	Reflex epilepsy
E	Myoclonic seizures	J	Psychogenic epilepsy
F	Atonic seizures		

Select one of the diagnoses above to fit each of the cases below.

1 A 23 year old who does not lose consciousness but has witnessed, brief, sudden drop attacks. They have been progressive and of longer duration more recently.

2 A 19 year old whose seizures are precipitated by the specific stimulus of flickering lights, as in certain television programmes.

3 Developing in childhood or adolescence, mostly after 5 years of age, seizures occur without warning/aura and the subject is often unaware that an attack has happened.

4

A	Paranoid	F	Schizotypal
B	Anankastic	G	Schizoid
C	Histrionic	H	Anxious-avoidant
D	Dependent	I	Dissocial
E	Borderline	J	Narcissistic

Which personality disorder is the individual described most likely to be diagnosed with?

1   A 32 year old with a strong sense of self-importance, shallow mood and a craving for excitement.

2   A 29-year-old man who interprets other people's actions as deliberately demeaning or threatening.

3   A 37 year old with an indifference to relationships and a restricted range of emotional experience and expression.

5

A	Coma	G	Leucocytosis
B	Type 2 diabetes	H	Hyperthermia
C	Proteinuria	I	Hypercalcaemia
D	Hypouricaemia	J	Rhabdomyolysis
E	Hyperphosphataemia	K	None of the above
F	Dehydration		

With regard to a 35-year-old man with neuroleptic malignant syndrome (NMS) treated with haloperidol:

1 Which of the above was a risk factor for NMS in this gentleman?

2 Which finding is common and diagnostically important?

3 Which finding is pathognomonic of NMS?

4 Which finding can lead to renal failure in NMS?

6

A   Over-inclusive thinking
B   Hierarchy of needs
C   Cohesiveness
D   Condensation
E   Loosening of associations

F   Concrete thinking
G   Drivelling
H   Abstract thinking
I   Depersonalisation
J   Primitive states of mind

Which of the above concepts are associated with the following figures?

1   Cameron

2   Bleuler

3   Schneider

4   Goldstein

5   Piaget

6   Bion

7   Yalom

7

A	Paranoid schizophrenia	F	Undifferentiated schizophrenia
B	Hebephrenic schizophrenia	G	Post-schizophrenic depression
C	Catatonic schizophrenia	H	Schizoid
D	Simple schizophrenia	I	Schizotypal
E	Residual/chronic schizophrenia	J	None of the above

Which diagnosis is being described in each case below?

1 Poor functioning and negative symptoms without preceding positive symptoms.

2 History of prominent persecutory, grandiose delusions and hallucinations present intermittently for 2 weeks.

3 A 25-year-old man with fleeting delusions, disorganised thought and rambling speech who was shy and solitary premorbidly.

8

A	Half-life	F	Bioavailability
B	First-order kinetics	G	Zero-order kinetics
C	Pure agonist	H	Inverse agonist
D	Antagonist	I	Second-order kinetics
E	Agonist	J	Pseudoreversible

The following descriptions relate to one of the pharmacological terms above.

1   Drug elimination is not proportional to the concentration of the drug in the body. The rate of drug elimination is constant.

2   The rate of drug elimination is directly proportional to its plasma concentration.

3   Maximal drug response at less than 100 per cent receptor occupancy.

9

A	Paranoid personality disorder	G	Anankastic personality disorder
B	Schizoid personality disorder		
C	Dissocial personality disorder	H	Anxious-avoidant personality disorder
D	Emotionally unstable – impulsive type		
E	Emotionally unstable – borderline type	I	Dependent personality disorder
F	Histrionic personality disorder		

Match each of the following descriptions to the most likely diagnosis.

1   A 32-year-old man who likes to be alone. Neighbours have been friendly, but he shuns them and avoids any conversation. He refuses to clear the rubbish from the front of his house, despite the council's demands.

2   A 24-year-old woman who can get angry quite easily. She is confrontational and quarrelsome, especially when criticised. Unstable and capricious mood.

3   A 22-year-old woman who feels socially inept, unappealing and inferior to others. Concerned about being criticised or rejected in social situations, she restricts her lifestyle because of a need for physical security.

10

A	Language	F	Memory
B	Object constancy	G	Frontal executive function
C	Prosopagnosia	H	Attention and concentration
D	Praxis	I	Visual agnosia
E	Primitive reflexes		

What is being tested or demonstrated in the following cognitive tests performed on a 73-year-old man with suspected degenerative dementia?

1 Copying the Rey–Osterreith figure.

2 An inability to name an object when pointed out to him.

3 Ask the patient to repeat a hand sequence of fist–edge–palm, five times with both hands.

INDIVIDUAL STATEMENTS: Answers

1   **False.** The role of D4 in relation to psychosis remains uncertain and D4 receptor antagonists have failed to show an antipsychotic effect. Serretti A et al. *Schizophr Bull* 1999, 25:609–18.

2   **False.** Amisulpride is a selective D2/D3 antagonist.

3   **False.** Bupropion (Zyban) is a nicotinic antagonist used in smoking cessation.

4   **True.** The Committee on Safety of Medicines advises that bupropion is avoided in patients with eating disorders, a history of seizures, central nervous system tumours and acute symptoms of alcohol or benzodiazepine withdrawal.

5   **False.** 30–60 per cent.
     Chong S-A. *Schizophr Bull* 2000, 26:421–40.

6   **True.**

7   **False.** Perfectionism is a robust and discriminating feature of anorexia nervosa.
     Halmi K et al. *Am J Psychiatry* 2000, 157:1799–805.

8   **False.** Secondary or long-term memory has a theoretically unlimited capacity.

9   **True.** The diagnosis and monitoring of anxiety in depressed patients are an important part of monitoring for suicidal risk.
     Schaffer A et al. *Can J Psychiatry* 2000, 45:822–6.

10   **False.** 10–25 per cent of patients with severe depression.

11   **False.** Suicide rates decline in wartime and at times of high employment. They increase during economic downturns and high unemployment.

12   **False.**
     Clarification involves rephrasing and questioning.
     Confrontation draws attention to what the patient is doing, e.g. missing appointments.
     Interpretation offers new formulations of unconscious meaning and motivation.

13 **True.** In depression, psychomotor retardation, delusions, history of previous response to ECT are predictors of a good response to ECT.

14 **True.** Others include fever, inflammation, head injury, hypercapnia and hypertension.

15 **True.**
Robinson T et al. *Arch Pediatr Adolesc Med* 2001, **155**:17–23.

16 **True.** More commonly bulimia nervosa.
Lacey JH, Moureli EJ. *Br J Addiction* 1986, **81**:389–93.

17 **False.** MDMA (3,4-methylenedioxymethamphetamine) is a hallucinogenic amphetamine. Phencyclidine and ketamine are dissociative anaesthetics.

18 **True.** PET studies have shown this.
McCann UD et al. *Lancet* 1998, **352**:1433–7.

19 **False.** Tetrachloroethylene is a volatile substance used in solvent cleaning products. It is commonly abused by teenagers. Others include toluene and butanes.

20 **False.** It accounts for 4 per cent. The National Confidential Inquiry into Suicide and Homicide in the UK has shown prescribed psychotropics to be the most commonly used.

21 **True.** A culture-bound syndrome of Latin America: 'If the soul leaves the body it becomes vulnerable to disease'.

22 **True.** Bulimic Investigatory Test Edinburgh (BITE).
Henderson M, Freeman CP. *Br J Psychiatry* 1987, **150**:18–24.

23 **True.**

24 **True.** It can (rarely) exacerbate schizophrenia and induce psychosis.

25 **False.** This is major depression superimposed on dysthymia. Treatment is with antidepressants.

26 **False.** It's the other way round.

27 **True.**

28 **True.** Associated with neuroleptic-induced extrapyramidal side effects. Often a late feature.

29   **True.** Parkinsonian tremor, muscular rigidity and akinesia are the cardinal features associated with neuroleptic-induced Parkinsonism (DSM-IV).

30   **False.** AIMS is a measure of neuroleptic-induced motor disturbance.

31   **True.** Flumazenil reverses the psychophysiological effects of the benzodiazepine agonists.

32   **False.** It reduces the levels of valproate, warfarin, clonazepam, dexamethasone, doxycycline, haloperidol. The levels of erythromycin are increased.

33   **True.** With subsequent reduced cerebral blood flow. There is a rebound increase in blood flow after withdrawal!

34   **True.** This has been shown in several studies looking at depression and sleep.

35   **False.** Veraguth described a triangle-shaped fold in the nasal aspect of the upper eyelid in patients with depression.

36   **False.** This was previously thought to be true, but all recent evidence suggests that there is no association.

37   **True.** Study involving 2007 patients with mania, 11 741 with depression, 81 380 with osteoarthritis and 69 149 with diabetes. However, depression may be an early symptom of dementia. Kessing LV, Nilsson FM. *J Affect Disord* 2003, **73**:261–9.

38   **False.** Tricyclic antidepressants delay cardiac conduction. There is a dose-dependent effect. This might precipitate complete heart block if there are pre-existing conduction delays.

39   **True.** Khat is a plant used recreationally mainly in East African countries, although its use is widely recognised in other parts of the world. Khat consumption has been associated with schizophreniform and manic psychoses.

40   **False.** The description is that of anosognosia. Somatopagnosia is the inability of a person to recognise one of their own body parts.

41   **True.** Memantine is thought to be an uncompetitive NMDA receptor antagonist, but its precise mechanism of action is complicated and currently uncertain.

42 **False.** Eicosapentanoic acid is a fish oil used to some effect in schizophrenia. It has a theoretical role in maintaining neuronal membrane structure.

43 **False.** Ninety per cent of patients with schizophrenia regularly smoke. Hypothetically, smoking may increase dopamine release, leading to a sense of well-being and reduction in negative symptoms.
Goff DC et al. *Am J Psychiatry* 1992, **149**:1189–94.

44 **True.** Drugs used include barbiturates (amobarbital, thiopentone), diazepam, midazolam. Abreaction is used to create a semi-conscious state, lowering the patient's defences and allowing them to speak about experiences or emotions. It is useful in distinguishing between dementia and pseudodementia or in hysterical disorders.

45 **True.** Sildenafil (Viagra) is also contraindicated in patients who have had a recent stroke, myocardial infarction or blood pressure < 90/50 mmHg.

46 **True.**

47 **True.** Trichotillomania (recurrent hair pulling) is also associated with infected follicles, trichophagy (hair swallowing) and trichobezoars (hair balls), which can lead to stomach pain and obstruction.
Christenson G et al. *J Clin Psychiatry* 1996, **57**:42–7.

48 **False.** It is greatest between 10 and 18 months. It usually disappears by 3 years.

49 **True.** An acetylcholinesterase inhibitor – not licensed in the UK.

50 **False.** Atasia abasia is the inability to stand or walk. It does not fit with organic disease and is often a symptom of conversion disorders.

51 **False.** Up to 90 per cent is protein bound.

52 **True.** Where a physiological reward mechanism following self-harm exists.

53 **True.** 1.9–3.1 per cent (Epidemiological Catchment Area Study).
Regier DA et al. *J Psychiatr Res.* 1990, **24 Suppl** 2:3–14.

54 **False.** Clomipramine is still used widely.

55 **False.** It increases slow-wave sleep. Benzodiazepines decrease slow-wave sleep and REM.

56 **True.** It also delays REM onset.

57 **True.** Freud distinguished between two types of object-choice, narcissistic and anaclitic. In narcissistic object-choice, the object is chosen on the basis of some real/imaginary resemblance to the self.

58 **False.** It is a subjective sense of a lump in the throat and difficulty swallowing with no evidence of abnormal anatomy.

59 **True.** Others include length of exposure to antipsychotic, increased age/elderly, female, head injury, organic brain disease and brain damage.

60 **False.** 121 million worldwide. 800 000 suicides/year; 450 million people with mental illness.
    WHO 2001, *Mental Health, New Hope* (World Health Report), Geneva, WHO.

61 **True.** Both are names for somatisation disorder.

62 **False.** The period is 2 years.

63 **True.** The risk of developing the condition was shown to rise by 30 per cent for every 10-year increase in paternal age. Study of 50 000 adolescent males.
    Zammit S et al. *Br J Psychiatry* 2003, 183:405–8.

64 **False.** It's the other way round.

65 **False.** Not really. In Alzheimer's disease there can be widespread slowing of the trace, slowing of alpha activity and non-characteristic changes; 95 per cent have a non-specific abnormal EEG.

66 **True.**

67 **True.** Neuroleptic-induced akathisia leads to restlessness and a sense of discomfort which can precipitate aggressive behaviour.

68 **False.** Duodenal dilatation is a complication of eating disorders and can be demonstrated by barium meal.

69 **False.** Model developed by Prochaska and DiClemente. Stages: 1. Precontemplative, 2. Contemplative, 3. Decision, 4. Action, 5. Maintenance, 6. Relapse.

70   **False.** Bipolar affective disorder is the most genetically determined disorder – heritability 0.8.

71   **True.** It involves 2 years of low mood where the diagnostic criteria for a depressive disorder are not met.

72   **False.** Lithium enters breast milk freely.

73   **True.**
      Wieck A et al. *BMJ* 1991, **303**:613–16.

74   **False.** Their risk is five times greater than that of the general population.

75   **False.** Weight (kg)/height (m^2).

76   **True.** Ratios of ≥ 1.0 for men and > 0.8 for women are associated with increased morbidity and mortality.

77   **True.** 'Acting in the best interests of the patient as stated in the Hippocratic oath'.

78   **True.** Excess salivation can lead to aspiration if the patient also has dysphagia.

79   **True.** However, not usually reaching diagnostic severity. Seasonal affective disorder is included in ICD-10 under recurrent depressive disorder.

80   **False.** Coexistence of these two conditions is common, especially vascular dementia.

81   **False.** Delayed grief is the onset of grieving 2 weeks after the event. Other abnormal reactions include inhibited grief, chronic grief and complicated grief.

82   **False.** 'Fatigue syndrome' is classified under Neurasthemia.

83   **False.** It is intermediate acting (6–20 hours). Midazolam is short acting and diazepam is long acting (50–100 hours).

84   **True.**

85   **True.**

86   **False.** This is covert conditioning. Counter-conditioning is when a conditioned stimulus is paired with a new stimulus that produces an incompatible or opposite response.

87 **True.** A desired low-frequency behaviour must be completed before a high-frequency behaviour can be carried out, e.g. you must complete 50 individual statements before you can watch TV.

88 **False.** The Rorschach test consists of ten ink blots and the subject is asked what each blot might represent. TAT is a test consisting of 20 cards of ambiguous scenes; the subject is asked to create a story about each card. Both are projective tests.

89 **True.** Speedballs comprise a stimulant and a 'depressant' used together intravenously. Heroin and cocaine are abused together in this way.

90 **True.** Although usually taken orally, it can also be absorbed transdermally, as eye drops, or intravenously.

91 **False.** Levels are low during the day and high during sleep. Melatonin is a hormone involved in circadian rhythms, which is produced by the pineal gland.

92 **True.** There are three main goals of cognitive therapy: to relieve symptoms and resolve problems, to acquire coping strategies, and to modify underlying cognitive structures in order to prevent relapse.

93 **True.** Acting out is a general character trait in which a person is given to relieving any intrapsychic tension by physical action.

94 **True.**
Hawton K, Fagg J. *Br J Psychiatry* 1988, **152**:359–66.

95 **True.** It may stimulate platelet aggregation and activation.
Wallaschofski H et al. *Horm Metab Res* 2004, **36**(1):1–6.

96 **True.**

97 **False.** Depression is the commonest, affecting up to 50 per cent of people with poorly controlled type 1 diabetes.
Pickup JC (ed.). *Brittle Diabetes*. Oxford, Blackwell Scientific, 1985, 76–102.

98 **False.** It lasts for 16 sessions, 24 for borderline personality disorder.

99 **True.** Others: deafness, dysphasia, dysarthria and learning disability.

100 **False.** It occurs in 1 per cent of treated patients.

101 **True.**

102 **False.** He described four basic operations of the mind (sensation, intuition, feeling and thinking) – which were considered with the two attitudes (introversion and extraversion) to create eight psychological types.

103 **False.** The figure is one-third.

104 **True.**

105 **False.** T4 and prolactin are unchanged; FSH and LH are reduced; growth hormone and cortisol are increased.

106 **True.** This is determined by the number of methyl groups attached to the nitrogen atom. A secondary amine has one group, a tertiary amine has two groups.

Tertiary amines (ADICT)	Secondary amines
Amitriptyline	Nortriptyline
Doxepin	Desipramine
Imipramine	Protriptyline
Clomipramine	
Trimipramine	

107 **False.** Tricyclic antidepressant plasma levels are reduced by anticonvulsants and increased by cimetidine, chlorpromazine and disulfiram.

108 **True.**

109 **False.** Sulpiride is pure D2. Amisulpride is D2/D3.

110 **False.** Neuroleptic malignant syndrome has been reported but is not common.
Umbricht D, Kane JM. *Schizophr Bull* 1996, **22**:475.

111 **True.**

112 **False.** 10 per cent.

113 **False.** They conform to the individual's idea of a disease.

114 **True.**

115 **True.** It is associated with progressive gait disorder, urinary incontinence and impaired mental function.

116 **False.** Features include wanting to get better, seeking professional advice, shedding some normal activities/responsibilities and exemption from blame.

117 **False.** These are command hallucinations, an important clinical feature.

118 **True.**

119 **False.** 15%. Easy 40%, Difficult 10%, Not defined 35%.
Thomas A, Chess S. *J Am Acad Child Psychiatry* 1977, 16(2):218–26.

120 **False.** Memory is relatively spared. Subcortical dementia, distractible, poorly developed EEG (loss of α activity), caudate and putamen atrophy.

121 **True.** Yes it can.

122 **False.** Married < never married < widowed < divorced.

123 **False.** It does not seem to change throughout life.
Kessing LV et al. *Psychol Med* 2003, 33(7):1177–84.

124 **True.**

125 **True.** Antidepressants have an opposite effect – down-regulating 5HT2 receptors.

126 **True.** It is used to control movement disorders, e.g. Huntington's chorea. It may act by depleting dopamine from nerve endings.

127 **True.** There is 15 per cent mortality over 20 years.

128 **False.** ECT increases the seizure threshold.

129 **False.** It is higher in men.

130 **True.** The lifetime prevalence of bipolar disorder is 1.2 per cent.

131 **False.** It is 0.1 per cent.

132 **True.**
Kay KC. *Clin Obstet Gynaecol* 1984, 11(3):759–86.

133 **True.**

EMIs: Answers

1

1   E

2   G

3   B

4   I

5   J

2

1   B   This is a common finding in conversion disorders. The patient's
          disorder does not fit the clinicians' understanding of disease-specific
          clinical signs.

2   I

3   D   Patients often perceive significant deformities which others consider
          mild or absent altogether.

3

1   F   The EEG is very variable and is normal between seizures.

2   I   Seizures can involve a motor component, e.g. sudden limb
          movements.

3   C   They can be provoked by hyperventilation. Subsequent confusion is
          uncommon and the disorder usually resolves by adulthood.

4

1   C

2   A

3   G

5

1   F   Other risk factors include:
        high-dose neuroleptics
        two or more neuroleptics
        depot medication
        previous NMS
        electrolyte disturbance
        male > female
        age < 20 years
        age > 60 years
        CNS dysfunction.

2   H   Recent neuroleptic exposure, hyperthermia, autonomic instability.

3   K

4   J   It is important to check for myoglobinuria.

6

1   A

2   D or E

3   G

4   F

5   H

6   J

7   C

7

1   **D**

2   **J**   The normal requirement for a diagnosis of schizophrenia (ICD-10) is that at least one of the diagnostic symptoms has been present for most of the time for at least 1 month.

3   **B**

8

1   **G**   Saturation kinetics: the concentration of the drug in the plasma may be high, but elimination is constant, limited, for example, by enzymatic availability.

2   **B**   So the greater the plasma concentration, the greater the amount of drug that is metabolised or removed (linear kinetics).

3   **C**

9

1   **B**

2   **D**

3   **H**

10

1   **F**   Visual memory (right temporal lobe).

2   **I**   Visual agnosia: unable to name an object by sight but often able to do so through another sense.

3   **G**   Luria's three-step task, a motor sequencing frontal lobe test.

PAPER 2

INDIVIDUAL STATEMENTS

1   The lack of a confiding relationship is a vulnerability factor for post-traumatic stress disorder.   T | F

2   The point prevalence of schizophrenia is 0.01 per cent.   T | F

3   The lifetime risk of depression is approximately 15 per cent for women.   T | F

4   An ambivalent relationship with 'a deceased' is associated with an abnormal grief reaction.   T | F

5   Digit symbol is a verbal assessment used in the Wechsler Adult Intelligence Scale.   T | F

6   Paramnesias are distorted recollections leading to memory falsifications.   T | F

7   Déjà entendu is the illusion of recognising a new thought.   T | F

8   Déjà pensé is the illusion of auditory recognition.   T | F

9   Klein believed that the ego and superego were present from birth.   T | F

10 Carbamazepine has a selective serotonin re-uptake inhibitor-like structure.   T | F

11 Behaviour therapy is the first-choice treatment for nocturnal enuresis in children.   T | F

12 Karl Abraham expanded Freud's oral and anal stages of development.   T | F

13 Thirty per cent of patients with a first depressive episode go on to have a relapse.   T | F

14 Thurstone described 17 primary abilities.   T | F

15 Winnicott is associated with the Scrabble game.   T | F

16 Object constancy is a component of Margaret Mahler's developmental theory.   T | F

17 Depersonalisation disorder is a remitting and relapsing disorder.   T | F

18	Chlorpromazine induces its own metabolism.	T	F
19	There is an association between leptin levels and weight gain with certain antipsychotics.	T	F
20	'Group think' is to be encouraged in the small group environment.	T	F
21	Moclobemide is selective for monoamine oxidase A.	T	F
22	Pseudohallucinations are under the individual's control.	T	F
23	Asyndesis is a feature of formal thought disorder.	T	F
24	Buspirone has a moderate sedative effect.	T	F
25	Pseudohallucinations are usually vivid experiences.	T	F
26	A reflex hallucination is a hallucination occurring in more than one sensory field.	T	F
27	Persisting with substance use despite clear evidence of overtly harmful consequences is one of the criteria for the dependence syndrome.	T	F
28	A dynamic formulation suggests a personal meaning of a person's symptoms in terms of his or her psychological organisation.	T	F
29	In dynamic terms, mania is considered to be a defence against anxiety.	T	F
30	One-third of people committing suicide have presented to the Accident and Emergency department on a previous occasion.	T	F
31	Narcotics Anonymous have a 12-step approach for problem drug users.	T	F
32	Menatetrenone (vitamin K2) may be beneficial in preventing bone loss in patients with anorexia nervosa.	T	F
33	Electric shock sensations are a feature of antidepressant discontinuation syndrome.	T	F
34	Yerkes–Dodson's theories of motivation relate to three drives – curiosity, competence and reciprocity.	T	F
35	Vitamin B12 deficiency is associated with anxiety and psychosis.	T	F
36	HLA DR2 has an association with sleep disorders.	T	F

37   A link between fluoxetine use and suicidal behaviour has been suggested.   T   F

38   The majority of people leave therapeutic communities in the first month.   T   F

39   Catharsis is purging of an emotion by experiencing it intensely.   T   F

40   Prolonged maternal separation perinatally leads to impaired attachment behaviour.   T   F

41   Interpersonal therapy has mainly been used for depression.   T   F

42   Socratic questioning is a core feature of dynamic psychotherapy.   T   F

43   The Beck Depression Inventory is a self-report measure.   T   F

44   Interpersonal therapy is a 12-session course of therapy.   T   F

45   Vicarious learning has been shown to lower anxiety.   T   F

46   Disgust is a secondary emotion.   T   F

47   Thurstone's primary mental abilities include memory.   T   F

48   Topiramate causes weight loss.   T   F

49   Positive reinforcement occurs when a behaviour is strengthened because an aversive stimulus has been escaped or avoided.   T   F

50   Splitting as a defence mechanism is associated with borderline personality disorder.   T   F

51   Free-radical-induced neurotoxicity has been suggested as a hypothesis explaining tardive dyskinesia.   T   F

52   Freud had five theories of anxiety.   T   F

53   Bleuler's criteria for schizophrenia include autism.   T   F

54   Pimozide is contraindicated in patients with QT prolongation.   T   F

55   The Tower of London Test is a test of parietal lobe function.   T   F

56   Narcolepsy is a dysomnia commonly associated with sleep paralysis and hypnagogic hallucinations.   T   F

57   Priapism is a common side effect of sildenafil.   T   F

58   There is a risk of contact sensitisation with chlorpromazine.   T   F

59   Agoraphobia is described as a fear of empty open spaces.   T | F

60   Carbon monoxide poisoning is a cause of catatonia.   T | F

61   According to the preparedness theory, things feared may have   T | F
     once been dangerous to the human race.

62   All dreams should be analysed in psychoanalysis.   T | F

63   Roth described phobic anxiety depersonalisation syndrome and   T | F
     considered it a form of anxiety neurosis.

64   Obsessive compulsive disorder is rarely associated with depressive   T | F
     illness.

65   Farmers have a low risk of suicide.   T | F

66   Nutmeg abuse is associated with psychosis.   T | F

67   Patients with depression who respond to acute antidepressants   T | F
     reduce the odds of relapse by 70 per cent if they continue
     treatment.

68   Risperidone reduces REM sleep in healthy individuals.   T | F

69   Morbid jealousy is a delusional state and not an overvalued idea.   T | F

70   During psychiatric interviewing, probing about suicidal intent   T | F
     probably increases the risk of suicide.

71   Visual hallucinations occur in oneiroid states.   T | F

72   Tactile hallucinations are a feature of cocaine withdrawal.   T | F

73   Bulimia has one of the highest mortality rates among psychiatric   T | F
     disorders.

74   Alcohol misuse has been shown to be one of the most consistent   T | F
     predictors of death among patients with anorexia nervosa.

75   Pentazocine can cause hallucinations.   T | F

76   There is a link between narcolepsy and HLA DR2.   T | F

77   A single use of cannabis will be detected in the urine for up to 27   T | F
     days.

78   Amphetamines are not detectable in the urine.   T | F

79   Morphine is a metabolite of heroin.   T | F

80	Diazepam is a metabolite of oxazepam.	T	F
81	Ego strength is a term coined by Bleuler.	T	F
82	Left–right disorientation occurs with dominant parietal lobe lesions.	T	F
83	There is loss of axillary and pubic hair in anorexia nervosa.	T	F
84	Frontal lobe functions can be impaired by maintenance ECT.	T	F
85	The risk of tardive dyskinesia reduces with increasing age.	T	F
86	Suicide has a significant association with obsessive–compulsive disorder.	T	F
87	Disclosure of information during interview is increased by commenting on interviewee affect.	T	F
88	The risk of completed suicides is higher in women than in men.	T	F
89	Asyndetic thinking is suggestive of schizophrenia.	T	F
90	There is a tendency to explore objects with the mouth in Klüver-Bucy syndrome.	T	F
91	Emergency surgery is a recognised precipitant of delirium tremens.	T	F
92	Gustatory hallucinations are suggestive of functional illness.	T	F
93	Pain is uninfluenced by cultural background.	T	F
94	Cimetidine inhibits metabolism of the benzodiazepines.	T	F
95	Two-thirds of people with Tourette's disorder meet diagnostic criteria for obsessive–compulsive disorder.	T	F
96	In obsessive–compulsive disorder, pathological thoughts are ego-syntonic.	T	F
97	Briquet's syndrome occurs more commonly in women.	T	F
98	Sighing is a symptom of anxiety.	T	F
99	The fear is often proportional to the threat in phobic states.	T	F
100	Tricyclic antidepressants have direct toxicity on the myocardium.	T	F
101	Edward and Gross are associated with the diagnostic criteria for benzodiazepine withdrawal.	T	F

102   A hysterical fugue has usually resolved by 3 weeks.   T | F

103   Ophthalmoplegia is a rare feature of Wernicke's encephalopathy.   T | F

104   Waxing and waning of the EEG is seen in drowsiness.   T | F

105   The alpha rhythm is 8–13 Hz in the normal EEG.   T | F

106   Onset before the age of 15 years is a good prognostic indicator in schizophrenia.   T | F

107   Negative symptoms in schizophrenia are associated with a poorer prognosis.   T | F

108   The risk of agranulocytosis with clozapine treatment is dose dependent.   T | F

109   The depressive position is associated with ambivalence towards the mother.   T | F

110   Modelling can result in social facilitation.   T | F

111   Defence mechanisms associated with hysteria include identification.   T | F

112   Visual hallucinations are common in schizophrenia.   T | F

113   Fluent speech is seen in nominal dysphasia.   T | F

114   Couvade syndrome and pseudocyesis are one and the same.   T | F

115   Orofacial dyskinesias are abnormal voluntary movements.   T | F

116   Propranolol is lipophylic.   T | F

117   Benzodiazepines cause retrograde amnesia.   T | F

118   DSM-IV makes use of a five-scale multi-axial system.   T | F

119   Depersonalisation and derealisation rarely occur together.   T | F

120   Dereistic thinking is the same as fantasy thinking.   T | F

121   Primary delusions are rarely preceded by a delusional mood.   T | F

122   Anhedonia is the complete loss of emotion.   T | F

123   Confabulation is a disorder of the temporal sequence of thinking.   T | F

124   Behavioural treatment of phobias is based on habituation.   T | F

125   Fear of blushing is a cognitive notion involved in cases of social   T | F
      phobia.

126   Imprinting is a psychological factor involved in simple phobias.   T | F

127   According to Beck's cognitive therapy, 'an individual's affect and   T | F
      behaviour are determined by the way he structures the world'.

128   Eidetic imagery involves memories stored with photographic-like   T | F
      quality.

129   The pre-operational stage of development is characterised by the   T | F
      ability to comply with rules.

130   Bottle-fed babies often show 'slow to warm up' personalities.   T | F

131   The transitional object is the 'good enough mother' according to   T | F
      Winnicott.

132   Depersonalisation should be treated in the main by ECT.   T | F

133   Beck's cognitive triad relates to positive thoughts about self, the   T | F
      present and the future.

EMIs

1

A Right frontal lobe
B Bilateral frontal lobes
C Left frontal lobe
D Dominant temporal lobe
E Non-dominant temporal lobe
F Bilateral temporal lobes
G Dominant parietal lobe
H Non-dominant parietal lobe
I Left occipital lobe
J Right occipital lobe
K Bilateral occipital lobes

Match the condition or physical sign to the affected brain area.

1 Wernicke's aphasia

2 Gerstmann's syndrome

3 Urinary incontinence

4 Agnosia for sounds and some qualities of music

5 Homonymous upper quadrantanopia

6 Anton's syndrome

7 Balint syndrome

2

A	Arbitrary inference	F	Minimisation
B	Selective abstraction	G	Catastrophising
C	Overgeneralisation	H	Jumping to conclusions
D	Dichotomous thinking	I	Personalisation
E	Magnification		

Match the histories provided to the cognitive distortions above.

1   A 35-year-old woman with high standards feels she is a complete failure if she is not completely successful in everything she does.

2   A 27-year-old psychiatric junior doctor gave a grand round presentation which was very well received. He feels it was a failure, however, as one slide had a spelling mistake on it.

3   A 22-year-old woman feels hopeless and has difficulty problem solving. She always thinks the worst: 'If it's going to happen, it's going to happen to me'.

3

A Regression
B Repression
C Reaction formation
D Isolation
E Undoing
F Projection
G Introjection

H Turning against the self
I Reversal
J Sublimation
K Idealisation
L Identification with the aggressor

Anna Freud described a number of defence mechanisms. Match the descriptions below to the correct defence mechanism.

1   Thinking of doing the opposite to the unconscious desire.

2   An ambivalently seen object is split in two, one object being seen as ideally good and the other as wholly bad.

3   The separation of an idea from its associated affect.

4   The indirect expression of instincts without adverse consequences.

4

A	Delusions of reference	F	Delusions of conspiracy
B	Delusional misinterpretation	G	Delusional memory
C	Delusional misidentification	H	Delusional jealousy
D	Delusional perception	I	Delusions of pregnancy
E	Delusions of persecution	J	Religious delusions

A patient presents to the Accident and Emergency department. He has a number of unusual beliefs, as shown below. Match them to the above psychopathological descriptions.

1   'People drop hints meant for me and say things with double meanings'.

2   'I see people that I recognise from a previous life'.

3   'My neighbour is trying to poison me by passing gas into my flat'.

4   'I see messages for me in the newspapers and on the television'.

5

A	Changing perceptions	F	Unfamiliarity
B	Dulled perception	G	Dysmorphophobia
C	Heightened perception	H	Déjà vu
D	Derealisation	I	Jamais vu
E	Depersonalisation		

A patient describes their symptoms to you. Choose the most appropriate definition.

1   'I feel as though I'm not a real person, not part of the living world'.

2   'I feel like I'm acting in a play with all the lines already written down for me'.

3   'Other people seem to be acting a part, like actors in a play or like puppets'.

4   'I feel unreal to myself when I look in a mirror or at a photograph'.

6

A	Likert scale	G	Schachter's cognitive labelling
B	Semantic differential		theory
C	Thurstone scale	H	Symbolic order
D	Pervasive communication	I	James Lange theory
E	Attribution theory	J	Cannon–Bard theory
F	Cognitive dissonance		

Which of the above is being described below?

1   Used to measure attitudes; presentation of a range of statements from which the subjects tick those they agree with.

2   A strive for consistency between cognitions thought to be related to each other.

3   Emotion derives from cortical perception and the physiological changes are hypothalamus driven and secondary.

7

A	Democratisation	F	Pairing
B	Introjection	G	Reactive motive
C	Division of labour	H	Double bind
D	Altruism	I	Reframing
E	Mirror phenomena		

Match the statements to the appropriate response above.

1   This forms part of Whitaker's 'focal conflict theory'.

2   This forms part of Bion's 'basic assumption theory'.

3   One of Yalom's therapeutic factors.

4   One of Foulke's 'group therapeutic factors'.

5   One of the four characteristics ideally in place in a therapeutic community according to Rapaport.

8

A	Universality	F	Socialising techniques
B	Altruism	G	Vicarious learning
C	Instillation of hope	H	Interpersonal learning
D	Imparting information	I	Group cohesiveness
E	Corrective recapitulation of primary family group	J	Catharsis
		K	Existential factors

Which of the above curative factors is described below?

1 Patients can feel improved or learn about themselves by helping fellow members.

2 The member realises that he is not unique in his awfulness.

3 Arousal of affect and expression by the patient.

9

A	Negativism	G	Jerkiness
B	Ambitendence	H	Freezing
C	Forced grasping	I	Automatic obedience
D	Echopraxia	J	Complex mannerisms
E	Flexibilitas cerea	K	Posturing
F	Opposition		

Which catatonic behaviour is described in each of the cases below?

1   A 37-year-old man keeps a leg in one position during voluntary movement.

2   A 56-year-old man is unable to complete a movement after seeming to try and again failing to complete.

3   In a 45-year-old woman, light pressure anywhere results in movement of her body in that direction.

10

A	Hyperthyroidism	F	Diabetes mellitus
B	Hypothyroidism	G	Phaeochromocytoma
C	Cushing's syndrome	H	Hypoglycaemia
D	Hyperparathyroidism	I	Wilson's disease
E	Hypoparathyroidism	J	Porphyria

Patients present with the following features; what is your diagnosis in each case?

1   A 42-year-old woman with an elevated serum calcium has low mood, reduced energy and motivation. Her husband first noticed a change in her mood 3 months ago.

2   A 36-year-old man presents to his general practitioner with incongruous behaviour and a change in personality. An MRI shows ventricular dilatation, brain-stem atrophy and basal ganglia hypodensities. Penicillamine is initiated following diagnosis.

3   A 55-year-old man with a history of 'attacks' which include headaches, perspiration, palpitations and marked anxiety.

INDIVIDUAL STATEMENTS: Answers

1 **False.** Brown and Harris's vulnerability factors relate to depression. (1) Lack of a confiding relationship. (2) Unemployment. (3) ≥ 3 children aged under 14 years of age living at home. (4) Loss of mother under age 11 years.

2 **False.** The prevalence is 1 per cent.

3 **True.** The risk is 9–26 per cent in women, 5–12 per cent in men.

4 **True.** It is also associated with dependent and insecure attachments.

5 **False.** *Verbal*: similarities, arithmetic, digit span, vocabulary, information and comprehension. *Performance*: object assembly, picture arrangement, block design, picture completion and digit symbol.

6 **True.** For example confabulation, déjà vu, déjà entendu, déjà pensé, jamais vu, retrospective falsification.

7 **False.** This is déjà pensé.

8 **False.** This is déjà entendu.

9 **True.**

10 **False.** It has a tricyclic antidepressant-like structure.

11 **True.**

12 **True.**

13 **False.** The figure is closer to 70 per cent.

14 **False.** He described seven: (1) verbal comprehension, (2) number, (3) word fluency, (4) inductive reasoning, (5) memory, (6) perceptual speed, (7) spatial orientation.

15 **False.** He is associated with the squiggle game. Also with transitional object, holding environment, potential space, at-one-ment, primary mental preoccupation, regression to dependence, going on being, object usage.

16 **True.** Normal autism 0–2 months, symbiosis 2–5 months, differentiation 5–10 months, practising 10–18 months, rapprochement 18–24 months, and object constancy 2–5 years.

17   **False.** It tends to be chronic and persistent.
      Baker D et al. *Br J Psychiatry* 2003, **182**:428–33.

18   **True.** As does carbamazepine, the phenothiazines and chloral
      hydrate.

19   **True.** Leptin may be associated with olanzapine and clozapine
      induced weight gain.
      Atmaca M et al. *J Clin Psychiatry* 2003, **64**:598–604.

20   **False.** The individual decisions of members of a group are
      suppressed in favour of the consensus. This can lead to flawed
      decision making.

21   **True.**

22   **False.**

23   **True.** It is lack of connection between thoughts.

24   **False.** It is rarely sedative.

25   **True.**

26   **False.** It is a hallucination associated with a stimulus in a different
      sensory modality.

27   **True.** ICD-10.

28   **True.**

29   **False.** It is a defence against depression – 'manic defence'.

30   **True.**
      Gairin I et al. *Br J Psychiatry* 2003, **183**:28–33.

31   **True.**

32   **True.** Vitamin K is necessary for the normal calcification of bone.
      Iketani T et al. *Psychiatry Res* 2003, **48**:259–69.

33   **True.** Also dizziness, low mood, insomnia, anxiety and
      agitation.

34   **False.** This is Bruner's theory. Yerkes–Dodson curve relates optimum
      performance with level of arousal.

35   **True.** It is associated with anxiety, depression, psychosis, acute
      organic reaction and dementia.

36   **True.** Almost all cases of narcolepsy are associated with HLA DR2.

37   **True.** The Committee on Safety of Medicines has advised that selective serotonin re-uptake inhibitors and venlafaxine should not be used in the under-18s to treat depression because of a potential increased risk of self-harm and suicide.

38   **True.** Up to 75 per cent leave in the first month.

39   **True.**

40   **False.** Bonding may be affected. Attachment up to 6 months is fairly indiscriminate.

41   **True.** It is usually 12–16 weekly sessions. It guides patients to understand current problems and helps them to find solutions. It also shares some features of dynamic psychotherapy.

42   **False.** This is the 'guided discovery' associated with cognitive behaviour therapy helping individuals challenge their thoughts or beliefs.

43   **True.**

44   **True.**

45   **True.**

46   **False.** Disgust is a primary emotion. Secondary emotions are combinations of primary emotions, e.g. love = joy + acceptance.

47   **True.** Memory, verbal comprehension, number, word fluency, inductive reasoning, perceptual speed and spatial orientation.

48   **True.** It is an anti-epileptic.

49   **False.** This is negative reinforcement.

50   **True.** Borderline – splitting into 'good' and 'bad' objects.

51   **True.** Theories include DA hypersensitivity, free-radical-induced neurotoxicity, GABA insufficiency and NA dysfunction.

52   **False.** He had three theories: (1) manifestations of repressed libido, (2) representation of the birth experience, (3) primary anxiety plus signal anxiety.

53   **True.**

54   **True.** It is an antipsychotic medication. An ECG should be done before administration.

55   **False.** It is a test of frontal lobe function.

56   **False.** Only 25 per cent have hypnagogic hallucinations.

57   **False.** It is rare. Others include dyspepsia, headache, flushing, dizziness and nasal congestion.

58   **True.** Care is needed if it is handled regularly.

59   **False.** It is a fear of crowds, public places, travelling away from home and travelling alone (ICD-10).

60   **True.**

61   **True.**

62   **False.**

63   **True.**

64   **False.** There is a 67 per cent lifetime risk of comorbid depressive illness.

65   **False.** Farmers have a high risk of suicide and suicidal thoughts. They have access to a number of potentially lethal methods. Thomas HV et al. *Occup Environ Med* 2003, **60**:181–6.

66   **True.** Nutmeg, a common household spice, is associated with psychosis if taken in excess. It is thought to have a similar chemical structure to serotonin antagonists such as reserpine. Kelly BD et al. *Schizophr Res* 2003, **60**(1):95–6.

67   **True.** Continuing antidepressant treatment reduces the odds of depressive relapse by around two-thirds. Geddes JR et al. *Lancet* 2003, **361**:653–61.

68   **True.** Sharpley AL et al. *J Clin Psychiatry* 2003, **64**:192–6.

69   **False.** Morbid jealousy can be a delusion or an overvalued idea.

70   **False.**

71   **True.** These are vivid elaborate scenic hallucinations occurring in schizophrenia and altered states of consciousness.

72   **False.** Cocaine 'bug' occurs in cocaine intoxication. Withdrawal involves dysphoria, anxiety, irritability, depression, fatigue and drug craving.

73   **False.** This is true of anorexia nervosa.

74   **True.**
        Keel PK et al. *Arch Gen Psychiatry* 2003, 60:179–83.

75   **True.** It is an opioid analgesic that can cause occasional hallucinations.

76   **True.**

77   **False.** Single use: 3 days; daily use: 10 days; heavy daily use: 21–27 days.

78   **False.** They are detectable in the urine for up to 2½ days.

79   **True.**

80   **False.** Oxazepam is a metabolite of diazepam.

81   **False.** The term was coined by Freud.

82   **True.** Dysphasia, dyslexia, constructional apraxia, visual disorientation and general intellectual impairment (plus Gerstmann's syndrome) also occur.

83   **False.** However, there is delayed secondary sexual development if pre-pubertal.

84   **True.** Maintenance ECT in this study was characterised by normal long-term memory but impaired short-term memory and frontal functions.
        Rami-Gonzalez L et al. *Psychol Med* 2003, 33:345–50.

85   **False.** There is a greater risk in the elderly.

86   **True.**

87   **True.**

88   **False.**

89   **True.** It involves lack of adequate connection between two consecutive thoughts.

90   **True.** The syndrome involves hyperorality, hypersexuality, hypermetamorphosis and hyperphagia.

91   **True.**

92   **False.** Gustatory and visual hallucinations are suggestive of organic illness.

93   **False.**

94   **True.**

95   **True.** Tourette's and obsessive–compulsive disorder have a similar age of onset and overlap in symptomatology.

96   **False.** They are ego-dystonic.

97   **True.** It involves multiple somatic symptoms in different areas.

98   **True.** The symptoms include shortness of breath, dizziness, syncope and paraesthesia, amongst others.

99   **False.** It is out of proportion.

100   **True.**

101   **False.** Associated with alcohol dependence (1976). They described compulsion, difficulties in controlling substance-taking behaviour, psychological withdrawal, tolerance, neglect of alternative leisure and persistence.

102   **True.**

103   **False.** This is a characteristic finding. There is also nystagmus, ataxia, clouding of consciousness and peripheral neuropathy.

104   **True.**

105   **True.**

106   **False.** It is associated with a poor prognosis.

107   **True.**

108   **False.**

109   **True.** This is a Kleinian concept. The infant (or patient in analysis) realises that both his love and hate are directed towards the same object – his mother.

110   **True.** It is associated with Albert Bandura. Positive outcome modelling is used in a variety of areas, including social skill development and parent training.

111   **True.** Denial, projection and identification.

112   **False.**

113   **True.** The patient has difficulty naming things, but his or her speech
        is fluent.

114   **False.** *Couvade syndrome*: the male complains of obstetric symptoms
        during his partner's pregnancy. *Pseudocyesis*: a false pregnancy
        affecting men or women.

115   **False.** These are involuntary movements.

116   **True.** Hence it crosses the blood–brain barrier readily. Propranolol
        causes nightmares. Atenolol is less lipid soluble.

117   **False.** They tend to cause anterograde amnesia. Benzodiazepine's
        mode of action is related to the opening of chloride channels
        facilitating the action of GABA.

118   **True.**

119   **False.** They commonly occur together.

120   **True.**

121   **False.**

122   **False.** Anhedonia is loss of the ability to experience pleasure.

123   **True.** Korsakoff commented on the 'confusion of temporal
        sequence of events and memories jumbled up and recalled
        inappropriately'.

124   **True.**

125   **True.**

126   **False.** It is the early rapid learning that allows a newborn or newly
        hatched animal to develop an attachment to its mother.

127   **True.** A person's structure of the world is based on cognitions that
        are based on assumptions or schemas developed from their previous
        experiences.

128   **True.**

129   **True.** Rules in this stage, according to Piaget, are sacrosanct: 'bad
        doing should be punished'.

130   **False.**

131   **False.** The transitional object is a toy/object to which the child is attached.

132   **False.**

133   **False.** This concerns 'negative thoughts'.

EMIs: Answers

1

1   **D**   Difficulty in comprehending the meaning of words.

2   **G**   Dyscalculia, agraphia, finger agnosia, right–left disorientation.

3   **B**

4   **E**

5   **D/E**

6   **K**   A failure to acknowledge blindness.

7   **K**   Triad of optic ataxia, oculomotor apraxia and simultanagnosia.

2

1   **D**

2   **B**   You could answer magnification to this question as well. Maximising the importance of a negative event.

3   **G**   Need to focus in therapy on the evidence that the worst does not happen.

3

1   **C**

2   **K**

3   **D**

4   **J**   For example anger is sublimated into painting.

4

1	A
2	C
3	E
4	A

5

1	E	
2	E	
3	D	Often depersonalisation and derealisation occur together.
4	E	Depersonalised perception of self.

6

1   C

2   F   Decrease dissonance by changing behaviour and dismissing
         information and by adding new thoughts to support one set of
         cognitions.

3   J

Note. Likert scale – agreement/disagreement on a five-point scale. Semantic differential – two adjectives/verbs at either end of a line which the subject marks (like visual analogue scale).

7

1 G The group organises around 'disturbing motive/avoided relationship', 'reactive motive/calamitous relationship'.

2 F Bion claimed primitive anxieties prevent groups working effectively through (i) pairing, (ii) fight/flight and (iii) dependence.

3 D Interpersonal input, catharsis, cohesiveness, self-understanding, interpersonal output, existential factors, universality, instillation of hope, altruism, family re-enactment, guidance and identification.

4 E Mirror phenomenon = group member like you.

5 A Democratisation, permissiveness, reality confrontation, communalism.

8

1 B

2 A

3 J

9

1 H

2 B

3 I

10

1   **D**

2   **I**   Penicillamine is a copper chelating agent used in the treatment of Wilson's disease.

3   **G**   Phaeochromocytoma is usually a benign tumour producing catecholamines. It causes periods of crisis lasting minutes with symptoms as described.

PAPER 3

INDIVIDUAL STATEMENTS

1. Academic under-achievement is associated with school refusal. T F

2. Discarding breast milk expressed up to 9 hours after a dose of sertraline reduces infant exposure by 17 per cent. T F

3. Depression is associated with a prolonged REM latency. T F

4. A shortened and lowered cortisol peak is associated with depression. T F

5. Higher order conditioning uses a conditioned stimulus as an unconditioned stimulus. T F

6. Children from lone parent families have been shown to be significantly more likely to suffer mental illness than other children. T F

7. The World Health Organisation '5 well being index' is a poor tool for detecting depression. T F

8. Outcomes in schizophrenia are worse in developing countries. T F

9. Myasthenia gravis often improves if the patient is taking lithium. T F

10. Morbid jealousy is more common in women than in men. T F

11. Learned helplessness is a concept derived from animal studies using escapable aversive stimuli. T F

12. Mirror gazing is a recognised feature of hebephrenic schizophrenia. T F

13. Failed habituation to glabellar tapping is an important clinical sign in Huntington's disease. T F

14. Carbamazepine can increase plasma haloperidol concentrations. T F

15. Lithium has been found to improve Parkinson's disease. T F

16. Lithium is associated with leucopenia. T F

17. Normal weight bulimia is associated with reduced sexual activity. T F

18   Muscarinic receptor availability is significantly reduced in   T | F
     untreated schizophrenic patients.

19   The prevalence of delusional disorders is greater in women than   T | F
     in men.

20   Patients with anorexia nervosa rarely show obsessional   T | F
     behaviours outside their eating disorder.

21   The male peak age of onset in schizophrenia is 5–10 years before   T | F
     the female peak.

22   Piaget's pre-operational stage is associated with animism.   T | F

23   Fears in the first year of life include a fear of animals.   T | F

24   Maternal deprivation is associated with infantile autism.   T | F

25   Depersonalisation is associated with the sense that the passage of T | F
     time has changed.

26   Motor clumsiness is common in Asperger's syndrome.   T | F

27   Post-traumatic stress disorder is not a condition of childhood.   T | F

28   Heavy smoking enhances clozapine metabolism.   T | F

29   Narcolepsy is usually associated with hypnagogic hallucinations.   T | F

30   The Stroop test assesses frontal lobe function.   T | F

31   Waxing and waning of the alpha rhythm (EEG) is associated with   T | F
     drowsiness.

32   Clozapine is associated with elevations in serum triglyceride.   T | F

33   Anxiety disorders rarely begin in old age.   T | F

34   Deliberate self-harm is commonest in the 22–28-year age range.   T | F

35   Amoxapine may act as an atypical antipsychotic.   T | F

36   The Hamilton cuff method was used in ECT to see if a seizure was   T | F
     occurring.

37   The overall lifetime prevalence of generalised anxiety disorder is T | F
     12 per cent.

38   Two-thirds of patients with obsessional disorders of recent onset T | F
     improve within a year.

39   Psychosis is the commonest psychiatric disturbance associated   T | F
     with Cushing's syndrome.

40   Clozapine treatment has shown potential for reducing suicidal   T | F
     behaviour in schizophrenia.

41   Neuroleptic malignant syndrome is a cause of delirium.   T | F

42   Cannabis confers a seven-fold increase in an individual's relative   T | F
     risk of subsequently developing schizophrenia.

43   The Cloniger type 2 alcohol dependent is usually less than 25   T | F
     years old.

44   Delirium tremens is associated with leucocytosis and deranged   T | F
     liver function.

45   A history of seizures is an indication for inpatient alcohol   T | F
     detoxification.

46   Carbon dioxide poisoning is a potential cause of Korsakoff's   T | F
     syndrome.

47   Quetiapine can be useful in the management of depressive   T | F
     symptoms in patients with psychosis.

48   Cognitive behaviour therapy involves time-limited (8–20), 1-hour   T | F
     weekly sessions.

49   Cognitive behaviour therapy is a problem-orientated treatment   T | F
     based on the 'now and then'.

50   Collaborative empiricism is part of the approach in cognitive   T | F
     behaviour therapy.

51   In psychotherapy, confrontation involves rephrasing and   T | F
     questioning.

52   Interpretation in psychotherapy offers new formulations of   T | F
     unconscious meaning and motivation.

53   In the psychotherapeutic process, clarification draws attention to   T | F
     what the patient is doing, sometimes on a regular basis.

54   Paranoid personality disorder is more common in males and   T | F
     people of lower social class.

55   Borderline personality disorder is more prevalent in younger age   T | F
     groups (19–34 years).

56 At 15-year follow-up, most subjects with borderline personality   T | F
    disorder no longer meet the criteria for the condition.

57 Compared to risperidone, olanzapine appears to be associated with   T | F
    a reduced risk of developing diabetes mellitus.

58 Transference is important in psychoanalysis.   T | F

59 Transference is overlooked in brief dynamic psychotherapy.   T | F

60 An ability to tolerate intimacy is a good prognostic indicator for   T | F
    analytic therapy.

61 Counter-transference is a hindrance in psychoanalytical   T | F
    treatment.

62 Family therapy is of use in the treatment of young people with   T | F
    anorexia.

63 In cognitive analytic therapy, the patient reads traps, dilemmas   T | F
    and snags in the file and identifies which apply to them.

64 A neurocognitive subtype of schizophrenia has been proposed   T | F
    based on impaired memory.

65 Operant conditioning is linked with Skinner and his rats.   T | F

66 Children observing an adult being aggressive with a large inflated   T | F
    doll are unlikely to mimic this behaviour.

67 A 4–9 month old uses one word, but is saying more than one   T | F
    word in meaning.

68 'Group think' occurs when members of a group are led to   T | F
    suppress their own dissent in the interests of the group consensus.

69 The 'Premack principle' holds that the frequency with which we   T | F
    engage in an activity is indicative of how rewarding it is.

70 In Erickson's psychosocial stages of development, intimacy versus   T | F
    stagnation represents early adulthood.

71 Anaclitic depression refers to the state into which an infant falls T | F
    following separation from a secure attachment.

72 Gestalt psychology is related to the preferential perception of   T | F
    organised wholes.

73 Short-term memory has a limited capacity of 5+/–2 chunks.   T | F

74 Episodic and semantic memory are examples of declarative memory. T | F

75 Peripheral neuropathy is not an essential feature of Korsakoff's syndrome. T | F

76 Anterograde amnesia is the memory loss for events occurring before the onset of a lesion. T | F

77 Verbal memory is particularly affected in the first 2–3 hours following ECT. T | F

78 Benzodiazepines impair learning and the acquisition of information in explicit tests of episodic memory. T | F

79 The Renfrew scheme tests expressive language. T | F

80 Night terrors occur in REM sleep, late in the night, and are remembered in the morning. T | F

81 A time-limited, structured, problem-orientated, collaborative process linking the present with the past describes cognitive behaviour therapy. T | F

82 The clearance of a drug is the rate at which the body eliminates a drug and is represented by the equation: $CL = 0.69 \times Vd/t^{1/2}$. T | F

83 Rawlins and Thompson classified adverse drug reactions. T | F

84 Normal intellectual functioning may be a protective factor against psychosis. T | F

85 Risperidone is associated with priapism. T | F

86 Major depression and dysthymic disorder affect 5–10 per cent of older adults in primary care. T | F

87 The suicide rate in Japan is one of the lowest in the world. T | F

88 Women with anorexia have lower than normal plasma leptin levels. T | F

89 The risk of tardive dyskinesia is increased in a patient who has other extrapyramidal side effects from antipsychotic medication. T | F

90 Approximately 25 per cent of patients with tardive dyskinesia will have a spontaneous remission. T | F

91	Almost 5 per cent of people who have a first episode of schizophrenia will not go on to have further episodes.	T	F
92	Risk genes have not been identified for schizophrenia.	T	F
93	Nefazodone inhibits the re-uptake of serotonin and also selectively blocks serotonin receptors.	T	F
94	Psychosis may accompany less severe depression.	T	F
95	Cannabis use among young people significantly increases their risk of developing depression and schizophrenia in later life.	T	F
96	Fifty per cent of extrapyramidal side effects appear within the first 5 days of commencing neuroleptic treatment.	T	F
97	Patients experiencing a first episode of schizophrenia can lose up to 13 per cent of grey matter during the first year of illness.	T	F
98	Dhat is considered an exotic 'neurosis of the orient'.	T	F
99	The traditional cut-off score for dementia, using the MMSE in mixed neuropsychiatric samples, is < 24/30.	T	F
100	Mentally ill people are six times more likely to be murdered than other individuals in the general population.	T	F
101	The risk of suicide among individuals with a family history of completed suicide is equivalent to that of people without a family history.	T	F
102	Olanzapine is associated with hyperlipidaemia in patients with schizophrenia.	T	F
103	The SCOFF questionnaire is used for the detection of eating disorders.	T	F
104	Reduced brain serotonin is found in impulsive bulimic individuals.	T	F
105	There is no current evidence for susceptibility loci for anorexia nervosa.	T	F
106	The risk of suicide for women with anorexia nervosa is ten times greater than that of healthy controls.	T	F
107	Carbamazepine blocks calcium channels, producing a reduced rate of recovery from excitatory stimuli.	T	F

108	Sodium valproate blocks GABA hydroxylase, thereby inhibiting GABA degeneration.	T	F
109	Cyclopyrolones act adjacent to the GABA complex.	T	F
110	Vigabatrin inhibits GABA-aminotransferase and is used to treat partial seizures.	T	F
111	Pindolol is a beta blocker and 5HT1a antagonist.	T	F
112	Donepezil is an irreversible inhibitor of central acetylcholinesterase.	T	F
113	Interferon treatment for hepatitis C is associated with depression.	T	F
114	A possible link between apolipoprotein D and psychosis has been demonstrated.	T	F
115	Milnacipran is an antidepressant with similar efficacy to imipramine.	T	F
116	Family intervention has little impact in reducing relapse and readmission rates in schizophrenia.	T	F
117	Borderline personality disorder cannot be treated with psychotherapy.	T	F
118	Spectinomycin increases blood levels of lithium.	T	F
119	Intermediate signs of lithium toxicity include chronic renal failure.	T	F
120	Pimozide is a thioxanthene.	T	F
121	Trazodone is a sedating antidepressant.	T	F
122	Mianserin can cause agranulocytosis.	T	F
123	Dextropropoxyphene leads to increased carbamazepine blood levels.	T	F
124	Treatment-resistant schizophrenia is defined as a failure to achieve acceptable remission of positive symptoms despite an adequate trial of two different classes of antipsychotic in adequate doses for at least 6 weeks.	T	F
125	Rabbit syndrome is a potential complication of phenothiazine usage.	T	F

126	Tiagabine only inhibits glial GABA uptake.	T	F
127	Bioavailability is the fraction of the dose which proceeds unaltered from the site of administration to the systemic circulation.	T	F
128	The Klüver-Bucy syndrome is associated with memory improvement.	T	F
129	The Klüver-Bucy syndrome is associated with blunting of emotions.	T	F
130	The Klüver-Bucy syndrome is associated with parietal lobe damage.	T	F
131	The Klüver-Bucy syndrome is associated with unrestrained exploring.	T	F
132	Secondary mania occurs with steroids.	T	F
133	Secondary mania has been reported with levodopa.	T	F

EMIs

1

A	Swing questions	G	Cognitive analytical therapy
B	Batch living requires core and cluster developments	H	Cognitive dysfunctions lead to a form of primitive thinking in depression
C	Recapitulations		
D	A division of depression into Types I and II	I	Obsessional phenomena
		J	Repressed memories of unhappy childhood experiences
E	Multi-thematic questions		
F	Total institutions are characterised by binary management		

1   Which of the above was suggested by Goffman?

2   Which of the above relates to Beck's formulation of depression?

3   Which of the above is a component of a good psychiatric interview technique?

2

A   Couvade syndrome   F   Paranoid personality disorder
B   Latah   G   Transsexualism
C   Dissocial personality   H   Post-traumatic stress disorder
D   Anankastic personality   I   Dissociative stupor
E   Munchausen's syndrome   J   Social phobia

Match each of the statements below to one of the above.

1   This disorder is associated with a fictional character who told wild stories.

2   This has an association with obsessive–compulsive disorder.

3   This is characteristically associated with little or no remorse.

4   This has a female prevalence of 1 in 100 000.

5   This disorder is associated with super-ego lacunae.

3

A	Paranoid schizoid position	F	Tangential thinking
B	Transference	G	Circumstantiality
C	Ego strength	H	Acting out
D	Operant conditioning	I	Time out
E	Agoraphobia	J	Super-ego

Match each of the descriptions below to the most suitable term above.

1   This arises from the resolution of the Oedipal complex.

2   This involves not being able to integrate the person as a whole object.

3   This often manifests with the ability to maintain long-term relationships.

4

A	Haloperidol	F	Flupentixol
B	Ibuprofen	G	Lamotrigine
C	Clozapine	H	Phenytoin
D	Sulpiride	I	Vigabatrin
E	Chlorpromazine	J	Mirtazapine

Answer the following.

1 Which of the above drugs increases lithium levels?

2 Which drug at low dose may have an alerting effect in schizophrenia?

3 Which drug is an example of a thioxanthene?

4 Which of these drugs inhibits GABA transaminase?

5

A	WAIS	F	Wechsler Preschool and Primary Scale of Intelligence (WPPSI)
B	National Adult Reading Test (NART)		
C	Mill Hill vocabulary test	G	Spearmans – (g)
D	Raven's progressive matrices	H	Stanford Binet
		I	Cattell
E	WISC-R	J	Hebb

Which of the above:

1   Is a general intelligence factor?

2   Is associated with two types of intelligence (A and B) – genetically based potential (A) and effective intelligence (B)?

3   Involves specifically, diagram completion?

4   Is used for preschool and primary school children (4–6½ years), including non-verbal and verbal components?

5   Is associated with 'fluid and crystallised ability'?

6

A	Citalopram	F	Nefazodone
B	Moclobemide	G	Milnacipran
C	Reboxetine	H	Ranitidine
D	Venlafaxine	I	Gabapentin
E	Mirtazapine	J	L-tryptophan

Which of the above drugs:

1   Is a selective noradrenergic re-uptake inhibitor?

2   Is a reversible inhibitor of monoamine oxidase type A?

3   Is a serotonin and noradrenaline re-uptake inhibitor?

4   Is a combined 5HT2 antagonist and 5HT re-uptake inhibitor?

5   Is a selective serotonin re-uptake inhibitor?

6   Is associated with eosinophilia–myalgia syndrome?

7   Has an isomer available which can be used to treat panic disorder?

7

A	Mydriasis	G	Anti-anxiety
B	Hyperphagia	H	Local anaesthetic
C	Hypersalivation	I	Antidepressant
D	Weight loss	J	Antipsychotic
E	Stimulant	K	Anti-obesity
F	Depressant		

This question relates to cocaine. Answer the following from the list above.

1   Is a feature of cocaine intoxication.

2   May occur with long-term cocaine usage.

3   A description of cocaine's central action.

4   An effect for which cocaine can be used clinically.

8

A	Alcohol	F	Cocaine
B	Nicotine	G	Cannabis
C	LSD	H	Benzodiazepines
D	PCP	I	Opiates
E	Methylenedioxymethamphetamine	J	Caffeine

Match each of the statements below to the most suitable answer above.

1   Is a serotonin analogue at 5HT2 receptors.

2   Increases intracellular concentration of cAMP and increases cerebral blood flow.

3   Causes a release of serotonin from the cortical system originating in the raphe nucleus.

4   Causes REM rebound and nightmares on withdrawal.

9

A	Pick's disease	F	Parkinsonism
B	Fragile X syndrome	G	Chronic fatigue syndrome
C	General learning disability	H	Asperger's syndrome
D	Nightmares	I	Agoraphobia
E	Mania	J	Late paraphrenia

Select from the above the condition that best fits each of the statements below.

1   Excess *Candida albicans* in the gastrointestinal tract has been implicated as a possible cause.

2   This can be diagnosed prenatally.

3   This is characteristically associated with stilted speech.

10

A	Alpha activity	F	Mu waves
B	Beta activity	G	V waves
C	Delta activity	H	Gamma activity
D	Theta activity	I	Spikes
E	Lambda waves	J	Sequential sharp waves

Which of the above EEG features relates to each of the descriptions below?

1　Frequency of between 8 and 13 Hz.

2　Occurs with ocular movements.

3　Attenuated with eye opening.

4　The predominant feature in 2–5 year olds.

5　Waveform activity increases with alcohol use.

INDIVIDUAL STATEMENTS: Answers

1   **False.**

2   **True.**
    Stowe ZN et al. *J Clin Psychiatry* 2003, **64**:73–80.

3   **False.** It is associated with shortened REM latency (the time between falling asleep and the first period of REM).

4   **False.** It is associated with a prolonged and raised cortisol peak.

5   **True.**

6   **True.** This was shown in a study looking at 65 000 single-parent children and 920 000 two-parent children in Stockholm. The relative risk of mental illness amongst the single-parent children was 2.1 for boys and 2.5 for girls.
    Weitoft GR et al. *Lancet* 2003, **361**:289–95.

7   **False.** WHO-5 was thought to improve family doctors' ability to detect depression.
    Henkel V et al. *BMJ* 2003, **326**:200–1.

8   **False.**

9   **False.** It worsens, especially muscle weakness.

10 **False.** It is more common in men – also called the Othello syndrome.

11 **False.** The studies used inescapable aversive stimuli.

12 **True.** The age of onset is usually 15–25 years.

13 **False.** It is a clinical sign in Parkinson's disease. The forehead is tapped with the examiner's finger and blinking habituation is observed.

14 **False.** It reduces plasma haloperidol levels.

15 **False.** Lithium can worsen Parkinson's disease.

16 **False.** It is associated with leucocytosis.

17 **False.** It is associated with normal or increased sexual activity.

18 **True.** SPECT studies have shown this, adding further evidence for the involvement of the muscarinic system in schizophrenia.
Raedler TJ et al. *Am J Psychiatry* 2003, **160**:118–27.

19 **False.**

20 **False.**

21 **True.** Peak ages of onset are 15–25 years (male) and 25–35 years (female).

22 **True.** They believe everything is 'living'.

23 **False.** Animal fear occurs at the age of 3–6 years.

24 **False.**

25 **True.** Many complain of a distortion in their perception of time and space.

26 **True.**

27 **False.** Cases do occur.

28 **True.**

29 **False.** Hypnagogic and hypnopompic hallucinations occur in approximately 25 per cent of patients with narcolepsy.

30 **True.** The Stroop test involves identifying the 'colour' words are printed in and not what they actually say. For example, if the word 'blue' is printed in the colour red, the correct response is red.

31 **True.**

32 **True.** It is often useful to measure the baseline lipid profile.

33 **True.** They usually develop at an earlier age.

34 **False.** It is commonest in the 15–24-year age range.

35   **True.** Amoxapine is a tricyclic antidepressant. It has a similar receptor occupancy profile to several of the newer atypical antipsychotics. Tardive dyskinesia is a recognised side effect.
Apiquian R et al. *Schizophr Res* 2003, 59:35–9.

36   **True.** A blood pressure cuff is inflated to above systolic pressure in one arm after the anaesthetic but before the muscle relaxant. Hence a seizure can be seen in the isolated limb.

37   **False.** The prevalence is about 5 per cent.

38   **True.** One-third run a chronic but fluctuating course.

39   **False.** Affective disorders are the most common. Psychiatric symptoms occur in 50–80 per cent of cases.

40   **True.** This was shown in a preliminary study of 980 patients with schizophrenia or schizotypal disorders.
Meltzer HY et al. *Arch Gen Psychiatry* 2003, 60:82–91.

41   **True.** Other potential causes include hypoxia, infection, metabolic disturbances, iatrogenic, vitamin deficiencies (thiamine, B12) and heavy metals.

42   **False.** It confers a twofold increase.
Arseneault L et al. *Br J Psychiatry* 2004, 184:110–17.

43   **True.** Cloniger Type 2 (male limited). Higher genetic component and antisocial behaviour. Usually less than 25 years. Impulsive and antisocial traits.

44   **True.** It is also associated with electrolyte imbalance and dehydration. The EEG shows an increase in fast activity.

45   **True.**

46   **False.** The cause is carbon monoxide poisoning.

47   **True.** Quetiapine was shown to be better than risperidone at reducing depressive symptoms, with fewer extrapyramidal side effects.
Sajatovic M et al. *J Clin Psychiatry* 2002, 63:1156–63.

48   **True.**

49   **False.** It is based on 'here and now'.

50 **True.** This is a systematic collection of information through self-monitoring, diaries and problem lists.

51 **False.** This is clarification.

52 **True.**

53 **False.** This is confrontation.

54 **True.**
Tyrer P, Duggan C, Coid C. *Br J Psychiatry* 2003, 182(Suppl. 44):s1–35.

55 **True.**
Tyrer P, Duggan C, Coid C. *Br J Psychiatry* 2003, 182(Suppl. 44):s1–35.

56 **True.**
Tyrer P, Duggan C, Coid C. *Br J Psychiatry* 2003, 182(Suppl. 44):s1–35.

57 **False.** Researchers looked at data from 19 153 patients receiving at least one prescription of olanzapine and 14 793 who received risperidone between January 1997 and December 1999; 319 patients taking olanzapine developed diabetes, compared to 217 taking risperidone.
Caro JJ et al. *J Clin Psychiatry* 2002, 63:1135–9.

58 **True.** Transference is the process by which the patient displaces feelings on to his or her therapist, re-enacting experiences from earlier influential relationships.

59 **False.** Freud originally found the transference process to be an irritant, only later believing it to be an essential part of therapy.

60 **True.**

61 **False.** Counter-transference refers to the therapist's attitudes towards the patient. It was also initially thought by Freud to be a hindrance.

62 **True.**

63 **True.** This type of therapy was developed by Anthony Ryle.

64   **True.** Patients with impaired memory exhibited more positive
      symptoms than their peers with good memory.
      McDermid Vaz SA, Heinrichs RW. *Psychiatry Res* 2002, 113: 93–105.

65   **True.** Skinner devised a chamber box in which a rat would have to
      press a lever to receive a reward.

66   **False.** They did mimic the behaviour. This was the experiment
      performed by Bandura in 1963 to demonstrate social learning theory.

67   **False.** This is the speech stage of a 9–18 month old (holophasic
      speech). A 4–9-month-old baby babbles, repeating syllables and
      making sounds.

68   **True.**
      Janis IL (1982). Groupthink Psychological studies of policy decisions
      and fiascos. Boston: Houghton Mifflin. They often arrive at poor
      decisions this way.

69   **True.** It is sometimes used as a part of the therapeutic approach in
      eating disorders.

70   **False.** Intimacy versus isolation represents early adulthood;
      generativity versus stagnation represents middle age.

71   **True.** The term was coined by Spitz to describe the state of protest,
      despair and detachment into which infants fall when separated from
      their mothers.
      Spitz RA. *Psychoanal Study Child* 1946, 2:313–342

72   **True.** It relates to the law of closure (seeing as a whole) and the law of
      continuity (see as a line).

73   **False.** It has a limited capacity of 7+/−2 chunks.

74   **True.** They are examples of declarative or explicit memory. Episodic is
      personal memories, e.g. autobiographical. Semantic memory is
      memory for knowledge and general information.

75   **True.**

76   **False.** This is retrograde amnesia. Anterograde amnesia is the
      impairment of learning new material.

77   **True.**

78   **True.** This can be reversed by benzodiazepine antagonists.

79   **True.** It involves naming multiple pictures or telling the story of a picture.

80   **False.** This is a description of nightmares. Night terrors occur as a rapid shift from stage 4 sleep to very light stage 1 sleep – they are not remembered by the child.

81   **False.** This describes cognitive analytical therapy as developed by Ryle.

82   **True.** Vd = volume of distribution; $t^{1/2}$ = half-life of the drug.

83   **True.** Type A: common, dose related, early onset. Type B: rare, dose independent, late onset.

84   **True.** Poor intellectual functioning might represent a vulnerability marker of brain dysfunction or an early stage of an evolving pathologic process that eventually leads to schizophrenia.
      Reichenberg A et al. *Am J Psychiatry* 2002, **159**:2027–35.

85   **True.** Risperidone is a potent α-adrenergic antagonist. This action may interfere with detumescence.
      Munarriz R et al. *N Engl J Med* 2002, **347**:1890–1.

86   **True.**
      Oxman TE et al. *Psychosomatics* 1990, **31**:174–80.

87   **False.** It is one of the highest – 25 per 100 000 population.
      Terao T et al. *Lancet* 2002, **360**:1892.

88   **True.**

89   **True.** It is also an increased risk with long-term high-dose neuroleptic treatment, in females, and in old age.

90   **True.**

91   **False.** The figure is about 30 per cent.

92   **False.** At least six susceptibility genes have been identified, which are thought to interact with environmental factors. In one example, researchers found that 21 per cent of a schizophrenia sample group had a polymorphism of the *Nogo* gene containing a CAA insert. This compared to just 3 per cent of the control samples.
      Novak G et al. *Mol Brain Res* 2002, **107**:183–9.

93   **True.**

94   **True.** In a study of 19 000 subjects, psychosis was often associated with severe depression, but also in up to 10 per cent of those with only two depressive symptoms. The authors suggest that psychosis may be missed in patients with mild to moderate depression.
Ohayon MM, Schatzberg AF. *Am J Psychiatry* 2002, **159**:1855–61.
Note. According to ICD-10, psychosis is only classified with severe depressive episodes if they meet the other criteria for depression.

95   **True.** Three studies published in the *BMJ* suggest that cannabis use among psychologically vulnerable adolescents represents a potentially serious risk to mental health.
Patton GC et al. *BMJ* 2002, **325**:1195–8.
Zammit S et al. *BMJ* 2002, **325**:1199–201.
Arseneault L et al. *BMJ* 2002, **325**:1212–13.

96   **False.** Ninety per cent appear within the first 5 days of treatment.
Casey DE. *Acta Psychiatr Scand* 1994, **89**(S380):14–20.

97   **False.** Three per cent of grey matter was lost in an MRI study of whole brain. The authors postulate that the decreases in grey matter volume may be explained by apoptosis and atrophy.
Cahn W et al. *Arch Gen Psychiatry* 2002, **59**:1002–10.

98   **True.** It is a culture-bound syndrome (semen loss anxiety).

99   **True.** MMSE is affected by age, educational level, cultural background, social class, literacy and language.

100   **True.** Records of the Danish Psychiatric Case Register were looked at. A quarter of all deaths among mentally ill patients were due to unnatural causes, with 1 per cent due to murder, 73 per cent to suicide and 26 per cent to accidental causes.
Hiroeh U et al. *Lancet* 2001, **358**:2110–12.

101   **False.** The risk of suicide among individuals with a family history was 2.58 compared with controls.
Oin P et al. *Lancet* 2002, **360**:1126–30.

102   **True.**
Koro CE et al. *Arch Gen Psychiatry* 2002, **59**:1021–6.

103   **True.** The SCOFF questionnaire is based on the following questions.
Do you make yourself sick because you feel uncomfortably full?
Do you worry you have lost control over how much you eat?

Have you recently lost more than one stone in a 3-month period?
Do you believe yourself to be fat when others say you are too thin?
Would you say that food dominates your life?
One point is given for every 'yes'; a score of 2 indicates a likely case of anorexia nervosa or bulimia.

104 **True.** Researchers believe that impulsivity is associated with alterations in 5HT mechanisms.
Steiger H et al. *Psychol Med* 2001, 31:85–95.

105 **False.** Chromosome 1p has been implicated in eating disorder susceptibility.
Grice DE et al. *Am J Hum Genet* 2002, **70**:787–92.

106 **False.** The risk is 57 times greater. (International Conference on Eating Disorders, Boston, USA, 2002.)

107 **False.** It blocks sodium channels.

108 **False.** GABA transaminase is blocked.

109 **True.** Examples include zopiclone, zolpidem.

110 **True.** It is used to treat partial seizures with or without secondary generalisation and is commonly associated with visual field defects as a side effect.

111 **True.** It has been used by some as an adjunct in refractory depression.

112 **False.** It is a reversible inhibitor.

113 **True.** Of 39 hepatitis C patients receiving interferon treatment, 33 per cent developed major depression.
Hauser P et al. *Mol Psychiatry* 2002, **7**:942–7.

114 **True.** Apolipoprotein D is involved in the binding and transport of cholesterol, steroid hormones (progesterone, pregnenolone, and arachidonic acid), which are all critical in brain growth and repair. Plasma ApoD levels were significantly higher in first episode psychotic patients compared to controls.
Mahadik SP et al. *Schizophr Res* 2002, **58**:55–62.

115 **True.** Milnacipran is a new antidepressant with essentially equal potency for inhibiting the re-uptake of both serotonin and noradrenaline.
Van Amerongen AP. *J Affect Disord* 2002, **72**:21–31.

116 **False.**
Leff J. *Acta Psychiatr Scand* 1994, **384**(Suppl.):133–9.
Lam D. *Psychol Med* 1991, **21**(2):423–41.

117 **False.**

118 **True.** It is a bactericidal antibiotic.

119 **False.** This is a late sign. Others include spasms, coma and fits.

120 **False.** It is a diphenylbutylpiperidine.

121 **True.**

122 **True.** A full blood count is recommended every 4 weeks for the first 3 months. It has fewer and milder antimuscarinic and cardiac side effects than amitriptyline.

123 **True.** It is a mild opioid analgesic, often combined with paracetamol as coproxamol.

124 **False.** Three different classes of antipsychotic are tried before being classified as treatment-resistant schizophrenia.

125 **True.** It is an oral dyskinesia.

126 **False.** It inhibits both neuronal and glial uptake. Tiagabine is used as an adjunctive treatment for partial seizures.

127 **True.**

128 **False.** It is associated with memory impairment.

129 **True.**

130 **False.** It is associated with bilateral temporal lobe impairment.

131 **True.**

132 **True.** Steroid treatment is associated with the risk of neuropsychiatric side effects, including psychosis.

133 **True.** Levodopa is the amino acid precursor of dopamine. It is used in the treatment of Parkinson's disease. Its side effects include involuntary movements and psychiatric symptoms, including psychosis and mania.

EMIs: Answers

1

1   F

2   H

3   C   Summarising and stating again the main points. Swing questions and questions on multiple themes are not considered best technique.

2

1   E   Baron von Munchausen.

2   D   Anankastic personality traits, not the personality disorder.

3   C   'Incapacity to experience guilt or to profit from experience'.

4   G   More commonly associated with male gender.

5   C   Super-ego: that part of the ego in which parental and societal values are incorporated.

3

1   J

2   A   According to Klein, the paranoid-schizoid position constitutes the infant trying to master its death instinct and precedes the depressive position.

3   C

4

1   **B**   Lithium levels are increased by: thiazide diuretics, NSAIDs, metronidazole, spectinomycin, acetylcholinesterase inhibitors, methyldopa.

2   **D**

3   **F**   Flupentixol, zuclopenthixol.

4   **I**   Vigabatrin, valproate.

5

1   **G**   Charles Spearman (1927) suggested that an individual's performance on a test of intellectual ability is determined by two factors: the g factor (a general factor) and the s factor (specific to a particular task).

2   **J**

3   **D**   They are supposedly less sensitive to cultural differences, but can also be used for those with communication difficulties.

4   **F**   Wechsler Preschool and Primary Scale of Intelligence (WPPSI).

5   **I**   Fluid intelligence is based on performance of relatively culture-free tasks. Crystallised intelligence depends on information already acquired, so is culture dependent.

6

1 C

2 B

3 D/G Milnacipran has been available in the USA for 6 years. It acts on noradrenaline and serotonin – almost equally (a 3:1 NA to serotonin balance). It is not yet available in the UK. In contrast, venlafaxine tends to act much more on serotonin than noradrenaline (a 1:30 NA to serotonin ratio).

4 F Nefazodone is now rarely used in the UK.

5 A

6 J Patients needed regular blood monitoring and to be registered with the Optimax Information and Clinical Support Unit (OPTICS) until 2005.

7 A Escitalopram.

Note. Mirtazapine is a presynaptic α_2-antagonist, increasing central noradrenergic and serotonergic neurotransmission.

7

1 A Others include:
    fever/chills
    sinus tachycardia/arrhythmia
    increased/low blood pressure
    dry mouth, anorexia, nausea and vomiting
    psychomotor agitation/retardation
    hyperactivity, irritability, aggressiveness
    convulsions, confusion, coma.

2 D Paranoid ideation, depression and weight loss.

3 E

4 H

8

1   **C**   Lysergic acid diethylamide (LSD). A hallucinogen.

2   **J**

3   **E**   MDMA or Ecstasy – a synthetic drug with both stimulant and hallucinogenic properties.

4   **H**

9

1   **G**

2   **B**   Also called Martin–Bell syndrome. It affects 1:1000 general male population and is associated with learning disability.

3   **H**   A pervasive developmental disorder.

10

1   **A**   Alpha (8–13 Hz), beta (> 13 Hz), delta (< 4 Hz), theta (4–8 Hz). Alpha activity is most prominent over the occipital region.

2   **E**

3   **A**   Alpha activity is attenuated with eye opening and accentuated by eye closure.

4   **D**

5   **A**

Note. Alpha activity is seen to diminish with age.

SECTION 2:
PART II MRCPsych

PAPER 4

INDIVIDUAL STATEMENTS

1   Secondary mania has been associated with bromide.   T F

2   In anorexia nervosa, superior mesenteric artery syndrome is a   T F
     medical complication.

3   Leucopenia with relative monocytosis is a recognised feature of   T F
     anorexia nervosa.

4   20 to 30 per cent of children with attention deficit hyperactivity   T F
     disorder do not respond to stimulant treatment.

5   Inhaled 35 per cent carbon dioxide has been used to distinguish   T F
     between generalised anxiety disorder and panic disorder.

6   Plasma homovanillic acid (pHVA) levels are reduced in prodromal   T F
     schizophrenic patients.

7   Persistent cognitive decline post coronary artery bypass grafting   T F
     is uncommon.

8   Individuals with depression have platelet abnormalities that may   T F
     predispose them to ischaemic heart disease.

9   In patients with coronary artery disease, symptoms of   T F
     anxiety and depression are predictors of poorer long-term
     functioning.

10   Patients treated for schizophrenia have an increased risk of   T F
     obstructive sleep apnoea.

11   P300 latency is increased in schizophrenia.   T F

12   Crowding in psychiatric wards is associated with increased   T F
     aggressive behaviour.

13   Incidence rates of hepatitis C are lower amongst individuals with   T F
     severe mental illness.

14   Amoxapine has both antidepressant and antipsychotic effects.   T F

15   Social rhythm disruption is associated with manic episodes.   T F

16   There is a high correlation between sex addiction and substance   T F
     misuse.

17   The blood oxygenation level dependent (BOLD) effect forms the   T   F
      basis of SPECT.

18   LAAM is used in the treatment of opioid dependence.   T   F

19   Depressed children have low growth hormone responses to GHRH   T   F
      compared to control subjects.

20   There is an increased prevalence of withdrawn behaviour,   T   F
      impaired social functioning and anxiety-related characteristics
      among relatives of autistic probands.

21   Institutional rearing protects children who are susceptible to a   T   F
      pattern of hyperactivity.

22   Specific reading retardation occurs with equal frequency in boys   T   F
      and girls.

23   Marchiafava-Bignami disease is characterised by visuospatial   T   F
      impairment.

24   Retrospective ruminative jealousy is a syndrome in which the   T   F
      patient is preoccupied with the past sexual activity of the present
      partner but without delusions of infidelity.

25   Cerebrospinal fluid tests for syphilis are positive in almost all   T   F
      patients with neurosyphilis.

26   Dialysis dementia occurs in 15 per cent of long-term dialysis   T   F
      patients.

27   In the UK a magistrate's intervention is needed prior to enforced   T   F
      nasogastric feeding in anorexic patients.

28   Suicide is more common in bipolar II patients than bipolar I   T   F
      patients.

29   Fifty per cent of patients with dysthymia also suffer recurrent   T   F
      major depression.

30   Anti-thyroid antibodies are present in 15 per cent of cases of   T   F
      depression.

31   L-Tryptophan is the precursor of the neurotransmitter   T   F
      noradrenaline.

32   Following ECT, post-synaptic β-adrenergic receptors are down-   T   F
      regulated.

33   Lithium potentiates the effects of the non-depolarising   T | F
      neuromuscular blockers, such as suxamethonium.

34   Stimulation of the amygdala causes placidity in animal studies.   T | F

35   Autoantibodies to serotonin are pathognomonic of   T | F
      fibromyalgia.

36   Amnesia is the rarest dissociative feature.   T | F

37   Multiple personality disorder is most common in late adolescence T | F
      and young adulthood.

38   The later the onset of multiple personality disorder, the worse the T | F
      prognosis.

39   An estimated 5000 people die each year worldwide through auto-   T | F
      erotic asphyxiation.

40   Memory impairment is the commonest cognitive complaint in   T | F
      patients with multiple sclerosis.

41   Flashbacks of hallucinations following hallucinogenic usage can   T | F
      occur for months afterwards.

42   An increase in the QTc interval of 30 ms from baseline after   T | F
      antipsychotic treatment is a definite cause for concern.

43   Lithium has been used to treat gout.   T | F

44   Seventy per cent of survivors of intracerebral haemorrhage have   T | F
      post-stroke depression.

45   Selective serotonin re-uptake inhibitors produce minor degrees of T | F
      tachycardia.

46   Chronic caffeinism starts at 1000 mg of caffeine per day.   T | F

47   ECT is best avoided in drug-resistant Parkinsonism.   T | F

48   Olanzapine is associated with increased leptin levels.   T | F

49   Maternal smoking during pregnancy has been associated with   T | F
      attention deficit hyperactivity disorder in offspring.

50   DSM-IV has criteria for the diagnosis of substance-induced   T | F
      persisting dementia.

51   A history of enuresis is associated with pyromania.   T | F

52 A history of malaria is a risk factor for the development of neuroleptic malignant syndrome. T F

53 First-born children have been found to have higher IQs than their younger siblings. T F

54 Vasopressin has been shown to stimulate appetite. T F

55 ANOVA is used to compare two or more groups of observations. T F

56 The non-acute porphyrias are not associated with neuropsychiatric complications. T F

57 Fasting can precipitate an episode of acute intermittent porphyria. T F

58 Sternbach's diagnostic criteria refer to the neuropsychiatric features of lupus. T F

59 Five per cent of women with a puerperal psychosis relapse with a subsequent childbirth. T F

60 There is a reduced response of growth hormone to clonidine in depression. T F

61 Cocaine increases slow-wave sleep and REM sleep. T F

62 Nightmares occur regularly in 1 per cent of the adult population. T F

63 Disulfiram treatment confers maximum benefit with regular supervision. T F

64 Acamprosate is a DOPA analogue used in the management of alcohol abstinence. T F

65 Tricyclics are second-line treatment in attention deficit hyperactivity disorder patients unresponsive to stimulants. T F

66 Dantrolene has been used successfully in the treatment of neuroleptic malignant syndrome. T F

67 The modern concept of autism derives from Kanner's description in the 1940s. T F

68 Klinefelter syndrome affects 1 in 700 live births and has the XYY karyotype. T F

69 Lesch Nyhan syndrome is due to a deficiency of hypoxanthineguanine phosphoribosyl transferase. T F

70   Patients with Diogenes syndrome have 14 days to appeal against   T | F
     Section 47 of the National Assistance Act 1948 (UK).

71   An enduring power of attorney is a legal device allowing   T | F
     individuals to assign an attorney to manage their finances in the
     event of them becoming mentally incapacitated.

72   A child of a patient with late-onset Alzheimer's disease has a risk   T | F
     of 1 in 5 to 1 in 6 of developing the illness.

73   Holders of a driving licence have a duty to disclose relevant   T | F
     mental illness lasting more than 3 months to the Driver and
     Vehicle Licensing Agency (DVLA).

74   Anti-Borna disease virus antibodies have been associated with   T | F
     mood disorders.

75   Insanity caused by intoxication or substance dependence can lead   T | F
     to a defence of insanity in court.

76   Frigophobia is the fear of corpses.   T | F

77   Cruelty to animals in childhood is a poor indicator of later   T | F
     violence against humans.

78   Phase IV of the clinical development stage of drug development is   T | F
     called post-marketing surveillance.

79   The Z test is a parametric test that examines the difference   T | F
     between an observed sample mean and a known population
     mean.

80   Logistic regression analysis cannot adjust confounders.   T | F

81   Naltrexone can be used to prevent relapse in patients with a   T | F
     history of opiate dependence.

82   Suicide is the leading cause of maternal death.   T | F

83   Standard deviation is equal to variance squared.   T | F

84   Statistical significance relates to events occurring by chance fewer   T | F
     than 1 in 20 times.

85   Posterior probability is the probability of an individual having an   T | F
     attribute after an event occurs.

86   Unrestricted random sampling is a technique that ensures an   T | F
     equal number of subjects is allocated to each group.

87 Bias is a systematic tendency that distorts inferred population values. T F

88 In an intention-to-treat analysis, trial dropouts are regarded as treatment failures and included in the analysis. T F

89 A distribution with its longer tail to the right is positively skewed. T F

90 A distribution with its longer tail to the left is negatively skewed. T F

91 A statistically significant test result is a value that is equal to or smaller than a chosen level of significance. T F

92 Power is the probability of not making a type I error. T F

93 In ecological studies, information about exposure and outcome is not gathered for each individual subject. T F

94 After two or more depressive episodes, the risk of further relapse approximates to 45 per cent. T F

95 At the age of 5 years, 10 per cent of children are wet at night. T F

96 Autogenic training is a machine-based technique that allows patients to detect related physiological signs of stress or anxiety and modify the signal and subsequently their behaviour. T F

97 Stimulant drugs can cause growth impairment. T F

98 Clonidine can reduce hyperactivity and impulsivity in autistic children. T F

99 Trichotillomania is an impulse control disorder. T F

100 In kleptomania, the urge to steal tends to be ego-dystonic, unlike the majority of shoplifters. T F

101 Topiramate is used in the treatment of bipolar disorders. T F

102 Topiramate causes increased appetite and weight gain. T F

103 A third of autistics are likely to develop epilepsy during their lives. T F

104 Erectile dysfunction occurs in 80 per cent of men with diabetes. T F

105 Eye movement desensitisation and reprocessing is used to treat panic disorder. T F

106 Personality disorders are estimated to occur in at least 10 per cent   T   F
of the population.

107 Lysergic acid is a 5HT2 agonist.   T   F

108 Psychic determinism relates to the assumption that our feelings,   T   F
behaviours, thoughts and symptoms are not random or arbitrary.

109 Vaginismus usually responds poorly to the available treatment   T   F
options.

110 A transient stress reaction can lead to episodes of shoplifting.   T   F

111 Approximately 25 per cent of patients receiving antipsychotics   T   F
have ECG abnormalities.

112 Parametric tests give better statistical power than non-parametric   T   F
tests for similar sample sizes.

113 Conduct disorders are associated with reading retardation.   T   F

114 Shoplifting is most common in females with previous   T   F
convictions.

115 Criminal responsibility is assumed in a child of 10 years of age.   T   F

116 Antidepressants increase EEG delta activity.   T   F

117 Childhood-onset schizophrenia is defined as onset of psychosis by   T   F
12 years of age.

118 An association between raised interleukin-2 activity and   T   F
depression has been shown.

119 In the EEG, α waves are most prominent over the frontal lobes.   T   F

120 20–25 per cent of patients with psychiatric diagnoses other than   T   F
schizophrenia have first-rank symptoms.

121 The social drift hypothesis suggests that the stresses related to   T   F
socioeconomic deprivation are risk-increasing factors for
schizophrenia.

122 Patients switched to clozapine from typical antipsychotics show a   T   F
reduction in basal ganglia volume.

123 Schizophrenic patients have impaired skin-conductance-   T   F
orientating responses to novel stimuli.

124 Epileptic twilight state is associated with marked perseveration of speech.   T | F

125 Sawtooth waves are associated with stage III sleep.   T | F

126 A cross-sectional study is a survey of the frequency of a disease or risk factor in a defined population at a given time.   T | F

127 Relative risk can be expressed as: incidence in exposed population/incidence in non-exposed population.   T | F

128 If the 95 per cent confidence interval for a treatment contains zero, the results are significantly different from no intervention.   T | F

129 The likelihood ratio for a test result compares the likelihood of that result in patients with disease to the likelihood of that result in patients without the disease.   T | F

130 A meta-analysis is a systematic review and summary of the medical literature.   T | F

131 The number needed to treat is the inverse of the absolute risk reduction.   T | F

132 Electromyogram burst amplitudes of less than 650 mV are associated with psychosis in Alzheimer's disease.   T | F

133 Ondansetron may be an effective monotherapy for obsessive–compulsive disorder.   T | F

134 Serious violence as an ictal phenomenon is common.   T | F

135 Exhibitionism can be classified into two groups.   T | F

136 Castration was banned from use in recidivist sex offenders in the UK in the 1950s after several high-profile cases.   T | F

137 Cyproterone acetate has been used for aggression in severe mental impairment.   T | F

138 There is an increased prevalence of schizophrenia in those convicted of homicide and other crimes of violence compared to the general population.   T | F

139 Anxiety states and neurotic depression are the rarest neuroses associated with offenders.   T | F

140 19 per cent of remand prisoners have a mental disorder.   T | F

141	The amygdala has a role to play in aggression.	T	F
142	Testamentary capacity requires knowledge of the nature and extent of one's property.	T	F
143	DSM-IV classifies bipolar I disorders as episodes of mania alternating with depression.	T	F
144	Women with recurrent depression are at risk of reduced bone density and osteoporosis.	T	F
145	Psychiatric illness causes or contributes to 12 per cent of all maternal deaths.	T	F
146	Untreated, the average duration of post-natal depression is 2 years.	T	F
147	A 2-week washout period is required when stopping fluoxetine for a monoamine oxidase inhibitor.	T	F
148	The incidence of obsessive–compulsive disorder is equal between the sexes.	T	F
149	The median age for a first depressive episode is 27 years.	T	F
150	The presence of aggression is associated with meaningful EEG changes.	T	F
151	Antichymotrypsin has been implicated as a possible susceptibility gene for Alzheimer's disease.	T	F
152	The amyloid β peptide linked to cognitive impairment in late-onset Alzheimer's disease appears to explain the cognitive impairment seen in schizophrenia.	T	F
153	Measles can lead to a progressive dementia.	T	F
154	Benzodiazepines act at specific benzodiazepine receptors.	T	F
155	Coprophagia is a feature of Gilles de la Tourette's syndrome.	T	F
156	Alpha lipoic acid is of benefit in Alzheimer-type dementia.	T	F
157	Excessive preoccupation with fantasy and introspection is consistent with a schizoid personality disorder.	T	F
158	Hypoplasia of the cerebellar vermis and cerebellar hemispheres has been described in autism.	T	F

159   Adverse parenting in childhood is associated with an increased   T | F
       risk of depressive disorders in adult life.

160   Depressed children are no more likely than controls to have non-   T | F
       depressive adult psychiatric disorders.

161   African-Caribbean people living in England are at higher risk of   T | F
       developing an illness that meets the operational criteria for
       schizophrenia than the populations in their countries of origin in
       the Caribbean.

162   The *MECP2* gene causes Rett's syndrome.   T | F

163   Eighty-two per cent of children in public care in the UK are of   T | F
       ethnic origin.

164   The drug treatment of choice for social phobia is a selective   T | I
       serotonin re-uptake inhibitor.

165   Family therapy is as effective as methadone in the treatment of   T | F
       opiate dependence.

EMIs

1

A	Analysis of variance	F	Kurtosis
B	Binomial distribution	G	Null hypothesis
C	Discriminant analysis	H	Receiver operating curve
D	F-test	I	Factor analysis of variance
E	Heterogeneity	J	Halo effect

Which of the above is being described below?

1   A test of three or more normally distributed independent groups compared on a single continuous measure.

2   A term used to describe the greater than expected difference in study results.

3   Represents a continuum of all the possible values of the cut-off point along the separator variable of a test.

2

A	Hawthorne effect	F	Significance bias
B	Obsequiousness bias	G	Publication bias
C	Sampling bias	H	Observer bias
D	Confounding	I	Recall bias
E	Residual confounding	J	Significance bias

Which of the above is being described in each of the following statements?

1 Participants improve their responses because they believe that interest has been taken in them.

2 Participants change their genuine responses to what they believe the researchers are interested in.

3 A factor which systematically affects all the dependent variables and is associated with the independent variables.

3

A	Stratification	F	Retrospective studies
B	Systematic sampling	G	Prospective studies
C	Observational studies	H	Experimental studies
D	Cross-sectional studies	I	Sequential studies
E	Longitudinal studies	J	Quasi-experimental studies

Match each of the descriptions below to one of the above.

1   In this type of study, each subject is observed once.

2   In this type of study, observations are collated with no artificial manipulation of important independent variables.

3   A study in which the information refers to outcomes occurring throughout the follow-up period.

4   Parallel and cross-over studies are examples of this type of study approach.

4

A   Dominant parietal lobe lesion
B   Dementia with Lewy bodies
C   Corpus callosum tumour
D   Delirium tremens
E   Non-dominant parietal lobe lesion

F   Wernicke's encephalopathy
G   Parkinson's disease
H   Prefrontal lobe tumour
I   Korsakoff syndrome

What is the most likely diagnosis in each case described?

1   A 67-year-old man who lives alone and is not known to psychiatric services attends the emergency room. He is dishevelled, agitated, sweaty, has a coarse tremor, and is tachycardic.

2   A 25-year-old man with a history of generalised convulsions and anosmia. On examination, he has a unilateral grasp reflex, facial and tongue weakness.

3   A 58-year-old man with unilateral inattention, dressing apraxia and astereognosis.

5

A	Thrombocytopenic purpura	F	Brushfield's spots
B	Microcephaly	G	Hydrocephalus
C	Blue eyes, fair complexion	H	Adenoma sebaceum
D	Underdeveloped secondary sexual characteristics	I	Hypogonadism
E	Macro-orchidism	J	Epilepsy

Which of the above are associated with each of the following conditions?

1   This feature is associated with Down's syndrome.

2   This feature is associated with XXY syndrome.

3   This is a feature of Martin–Bell syndrome.

4   This is a feature of Prader–Labhart–Willi syndrome.

5   This is found in 50 per cent of children with tuberous sclerosis by the age of 5 years.

6   This is commonly found in phenylketonuria.

6

A	Acetylcholine	F	Glycine
B	Noradrenaline	G	Histamine
C	Serotonin	H	Dopamine
D	Glutamate	I	Monoamine oxidase
E	GABA	J	N-methyl-D-aspartate

1 The nucleus basalis of Meynert is associated primarily with which of the above?

2 The locus ceruleus is associated primarily with which of the above?

3 The raphe nucleus is associated primarily with which of the above?

4 The release of which of the above is inhibited by lamotrigine?

INDIVIDUAL STATEMENTS: Answers

1   **True.**

2   **True.** Superior mesenteric artery syndrome is an uncommon but well-recognised clinical entity characterised by compression of the third, or transverse, portion of the duodenum against the aorta by the superior mesenteric artery, resulting in complete or partial duodenal obstruction.

3   **False.** Leucopenia with relative lymphocytosis is recognised.

4   **True.** Due to medication inefficacy or unacceptable adverse effects. Connor D et al. *Clin Pediatr* 2000, **39**:15–25.

5   **True.** Patients with generalised anxiety disorder do not react adversely to breathing 35 per cent carbon dioxide, unlike patients with panic disorder.
Perna G et al. *J Clin Psychiatry* 1999, **60**:379–84.

6   **False.** Levels of pHVA (the major metabolite of dopamine) are an indicator of central dopamine activity. Prodromal patients have been shown to have elevated levels of pHVA compared to controls.
Sumiyoshi T et al. *Biol Psychiatry* 2000, **47**:428–33.

7   **False.** Cognitive decline following coronary artery bypass grafting is common and is persistent in many patients.
Newman N et al. *N Engl J Med* 2001, **344**:395–402.

8   **True.** Studies have shown that:
psychological stress can influence the balance between formation and dissolution of thrombi
platelet activation is altered in depression
plasma levels of thrombus-promoting factors are significantly increased in elderly patients with depression.
Nemeroff C. *Am Heart J* 2000, **140**:57S–62S.

9   **True.** Depression and anxiety were more predictive of functional status than coronary arteriography at 1 year.
Sullivan M et al. *Am J Cardiol* 2000, **86**:1135–8.

10 **True.** Long-term neuroleptic use can lead to obesity, which is a risk factor for obstructive sleep apnoea.
Winkelman J. *J Clin Psychiatry* 2001, **62**:8–11.

11 **True.** This is a cognitive event-related potential shown to have reduced amplitude and prolonged latency in schizophrenia.

12 **True.** Especially verbal aggression
Ng B et al. *Psychiatry Serv* 2001, **52**:521–5.

13 **False.** There are higher incidence rates of hepatitis C and B and HIV in individuals with severe mental illness.
Rosenberg S et al. *Am J Public Health* 2001, **91**:31–7.

14 **True.** It is a tricyclic antidepressant related to the antipsychotic loxapine.
Vega W et al. *J Clin Psychopharmacol* 2000, **20**:504–19.

15 **True.** Social rhythm disruptions are alterations in routine such as eating, sleeping and exercise, and may be associated with the initiation of affective episodes.
Malkoff-Schwartz S et al. *Psychol Med* 2000, **30**:1005–16.

16 **True.** A correlation has been shown of up to 80 per cent in some studies. Sexual addiction is not a diagnosable disorder in DSM-IV or ICD-10, but has developed as a concept over the last 20 years.

17 **False.** BOLD effect is acquired in fMRI.

18 **True.** Levo-acetyl alpha methadol (LAAM) has been used with some efficacy.

19 **True.** This is a finding replicated in many studies. This may represent a trait marker.
Dahl R et al. *Biol Psychiatry* 2000, **48**:981–8.

20 **True.** A study looking at 99 probands with autism found particular traits may aggregate among relatives of autistic individuals.
Murphy M et al. *Psychol Med* 2000, **30**:1411–24.

21 **False.**
Roy P et al. *J Child Psychol Psychiatry* 2000, **41**:139–49.

22 **False.** It occurs with greater frequency in boys.

23 **False.** Marchiafava-Bignami disease is characterised by ataxia, epilepsy, dysarthria and impaired consciousness. It is a slowly progressive form of dementia associated with heavy alcohol consumption. Spastic paresis also occurs.

24 **True.**

25 **True.**

26 **False.** It has become much rarer with the introduction of aluminium-free dialysates.

27 **False.** The House of Lords ruled that nasogastric feeding in eating disorders constituted treatment of a psychiatric disorder and as such can be used (Riverside Health NHS Trust v. Fox 1994).

28 **True.** *Bipolar I*: episodes of depression and mania.
Bipolar II: episodes of depression and hypomania.

29 **True.** The Epidemiological Catchment Area Study showed this (1988).

30 **True.** They are present in 9–20 per cent of cases.

31 **False.** L-Tryptophan is an essential amino acid and the precursor of serotonin.

32 **True.** This also happens with antidepressant treatment. ECT affects all the neurotransmitter systems in some way.

33 **True.** Hence care should be taken in patients receiving ECT who are also taking lithium. Also, lithium may lower the seizure threshold and lead to prolonged seizures.

34 **False.** Studies have shown that stimulation of the amygdala causes aggression and ablation leads to placidity.

35 **False.** Serotonin antibodies are widely found in the population. They might be increased in patients with certain psychiatric conditions such as schizoaffective psychosis.
Schoff K et al. *Psychiatry Res* 2003, **121**(1):51–7.

36 **False.** Amnesia is the commonest symptom and occurs in almost all of the dissociative disorders.

37 **True.** Multiple personality disorder or dissociative identity disorder is also more common in women (90–100 per cent of cases).

38 **False.** The earlier the onset of multiple personality disorder, the worse the prognosis. Much debate exists as to the authenticity of multiple personality disorder. It is very difficult to treat.

39   Fa|se. The figure is 50–100 people per year. Mild hypoxia is purported to heighten orgasm intensity.

40   **True.** It occurs in 40–60 per cent of patients.

41   **True.** DSM-IV recognises a syndrome called hallucinogen persisting perception disorder.

42   **False.** Although this would be of some concern, a 60 ms QTc increase is a definite cause for concern. QT interval prolongation is a risk factor for torsade de points, which is associated with sudden death.
Bednar RR et al. *Prog Cardiovasc Dis* 2001, 43:1–45.

43   **True.** It was used for this purpose in the 1800s.

44   **False.** 30–40 per cent of survivors of intracerebral haemorrhage have post-stroke depression.
Katona C, Livingston G. *Lancet* 2000, **356**:91–2.

45   **False.** Selective serotonin re-uptake inhibitors tend to cause minor bradycardia.
Roose SP et al. *Arch Gen Psychiatry* 1987, **44**:273–5.

46   **False.** This is considered toxic caffeinism. Chronic caffeinism is between 600 and 750 mg of caffeine per day.

47   **False.** ECT can be used in drug-resistant Parkinsonism.
Anderson K et al. *Acta Neurol Scand* 1987, **76**:191–9.

48   **True.** Leptin is produced by fat cells. Olanzapine-associated weight gain may be related to increased leptin levels.

49   **True.** Maternal smoking during pregnancy remained a significant influence when other potential confounders were taken into account.
Thapar A et al. *Am J Psychiatry* 2003, **160**:1985–9.

50   **True.** These are coded according to the substance involved, e.g. alcohol, inhalant, sedatives, hypnotics.

51   **True.** Others include poor academic performance, learning disability, attention deficit hyperactivity disorder, speech problems, cruelty to animals, visual and other physical defects.

52   **False.** Risk factors include high-dose neuroleptic medication, rapid increase in dose, parenteral/depot medication, organic brain disease,

Parkinson's disease, dehydration, electrolyte disturbance, psychomotor agitation, history of catatonia and previous neuroleptic malignant syndrome.

53 **True.** This is possibly due to greater parental interaction with the first child.

54 **False.** Vasopressin (and oxytocin) are involved in the regulation of mood. Both are synthesised in the hypothalamus and released in the posterior pituitary.

55 **True.**

56 **True.** Unlike the acute porphyrias such as acute intermittent porphyria, variegate porphyria, hereditary coproporphyria.

57 **True.** It is also caused by drugs (e.g. barbiturates, sulphonamides, oral contraceptive pill), acute infections and alcohol.

58 **False.** They refer to the serotonin syndrome (serotonin hyperstimulation). At least three of the following should be present for a definite diagnosis: agitation/restlessness, sweating, diarrhoea, fever, hyperreflexia, lack of coordination, mental state changes (confusion, hypomania), myoclonus, shivering, tremor.

59 **False.** The figure is 20 per cent.

60 **True.**

61 **False.** Cocaine reduces slow-wave sleep and REM. Excessive sleep occurs on withdrawal.

62 **True.**

63 **True.** Disulfiram is an irreversible inhibitor of aldehyde nicotinamide adenine dinucleotide (NAD) reductase. It increases acetaldehyde from incomplete alcohol metabolism, leading to a flushed face, tachycardia, nausea, vomiting and hypotension.
Hughes JC, Cook CC. *Addiction* 1997, 92:381–95.

64 **False.** Acamprosate is a GABA analogue. It has a possible anti-craving effect. It is licensed for 1 year's treatment in the UK.

65 **True.** In specialist settings, imipramine, clomipramine, nortriptyline and desipramine are sometimes used.

66 **True.** Dantrolene is a skeletal muscle relaxant, reducing duration and mortality in neuroleptic malignant syndrome. Other drugs include amantadine, bromocriptine and benzodiazepines.

67 **True.** 'A disorder characterised by delayed and abnormal development of social relationships and language'.

68 **False.** XXY = Klinefelter syndrome. It affects 1 in 500 births. Affected males lack secondary sexual characteristics and are infertile. They have behavioural difficulties, psychiatric disorders and low IQ. XYY occurs 1 in 700 births. Affected males are tall and have behavioural problems.

69 **True.** It leads to hyperuricaemia. Clinical features are choreoathetosis and self-mutilation (commonly of the lips and face).

70 **False.** Diogenes syndrome = senile squalor syndrome. This act allows the person to be removed from their home with **no** right of appeal.

71 **True.**

72 **True.**
Liddell M et al. *Br J Psychiatry* 2001, **178**(1):7–11.

73 **True.**

74 **True.**
Hayato T et al. *Psychiatry Res* 2003, **120**:201–6.

75 **True.**

76 **False.** It is a culture-bound syndrome associated with fear of the cold, loss of energy and compulsive wearing of many layers of clothing.

77 **False.** Childhood cruelty to animals has predictive value for later aggression against humans.
Felthous A et al. *Am J Psychiatry* 1987, **144**:710–17.

78 **True.**
Phase 1: normal volunteers give pharmacodynamic data etc.
Phase 2: patients confirm pharmacodynamic data etc.
Phase 3: patients used in clinical trials.
Phase 4: post-marketing surveillance.
Phase 5: ongoing trials.

79 **True.**

80 **False.** It can be, and is, used to adjust for confounders.

81 **True.** It is an opioid antagonist, given to former addicts in an attempt to prevent relapse.

82 **True.**
1997–1999 Report of the Confidential Enquiries into Maternal Deaths (2001).

83 **False.** SD = √variance.

84 **True.** < 5 per cent or $p < 0.05$.

85 **True.** Prior probability is the probability of an individual having an attribute before an event occurs.

86 **False.** This is restricted randomisation. In unrestricted randomisation and sampling, selection is with replacement; each member of a population has a calculable, known and equal probability of being selected.

87 **True.**

88 **True.**

89 **True.**

90 **True.** Skewness describes the symmetry of a distribution with respect to its mean.

91 **True.** Usually, by convention, $\alpha = 0.05$ (5 per cent).

92 **False.** Power is the probability of not making a type II error, i.e. the probability of rejecting a false null hypothesis.

93 **True.** Ecological studies are used to generate a hypothesis between an exposure and outcome based on aggregated data.

94 **False.** There is > 90 per cent risk; 50 per cent after one episode.

95 **True.** At 5 years, 3 per cent of children are wet during the day.

96 **False.** Biofeedback is described. Autogenic training is a series of structured suggestions that promote body suggestions associated with relaxing.

97 **True.** For example Ritalin in children. It is important to give regular drug holidays and to chart their growth.

98 **True.** It is thought to have some effect in reducing hyperkinetic behaviour.

99 **True.** It is compulsive pulling of one's own hair. Other impulsive disorders include intermittent explosive disorder, pyromania, kleptomania and compulsive gambling.

100 **True.**

101 **True.** Although it is currently not a first-line treatment.

102 **False.** It causes anorexia and weight loss.

103 **True.**

104 **False.** The figure is 50 per cent.
McCulloch DK et al. *Diabetologia* 1980, 18(4):279–83.

105 **True.** Eye movement desensitisation and reprocessing, developed by Francine Shapiro, was initially used for post-traumatic stress disorder but is now also used for other anxiety disorders.

106 **True.**
Oldman J. *JAMA* 1994, **272**:1770–6.

107 **True.** This is LSD.

108 **True.**

109 **False.** Specific treatment techniques and exercises are extremely effective.

110 **True.** This is reactive shoplifting.

111 **True.**
Thomas S. *Adverse Drug React Toxicol Rev* 1994, **13**:77–102.

112 **True.**

113 **True.**

114 **True.**

115 **True.** In the UK, 10–14 year olds are assumed to know the difference between right and wrong (e.g. Thomson and Venables, the Jamie Bulger case).

116 **True.** They also increase beta and theta activity.

117 **True.** It is rare, and usually severe, and has been shown to be continuous with the adult disorder.

118 **True.**

119 **False.** They are most prominent over the occiput: accentuated by eye closure, attenuated by eye opening.

120 **True.**

121 **False.** This is the social-causation hypothesis. The social drift hypothesis suggests that individuals drift down to lower socioeconomic classes as a consequence of social and occupational incompetence associated with schizophrenia.

122 **True.** It may be due to clozapine's low potential for extrapyramidal side effects and tardive dyskinesia.
Chakos MH et al. *Lancet* 1995, 345:456–7.

123 **True.** About 50 per cent fail to produce it and others fail to habituate.

124 **True.** Epileptic twilight state: impaired consciousness, several hours (rarely > 1 week), visual hallucinations, abnormal affective states (panic, terror, anger, ecstasy), psychomotor retardation with marked perseveration of speech.

125 **False.** Stage V: REM/sawtooth. Stage I: theta waves. Stage II: sleep spindles and K complexes. Stage III: delta waves, 20–50 per cent. Stage IV: delta > 50 per cent.

126 **True.**

127 **True.**

128 **False.** Results are not significantly different from 'no treatment'.

129 **True.**

130 False. This is an overview. A meta-analysis is an overview that uses quantitative methods to summarise the results.

131 **True.**

132 **True.** They occurred in 69 per cent of psychotic Alzheimer's disease patients compared to 17 per cent without psychosis. The authors suggest it may help identify individuals susceptible to developing psychosis.
Caliguiri MP et al. *Neurology* 2003, **61**:954–8.

133 **True.** 5HT3 antagonists have been found to act as anxiolytics in selected animal models of anxiety and a small sample of obsessive–compulsive disorder patients. It is currently licensed for use in peri-operative nausea or that associated with chemotherapy.

134 **False.** It is very rare.

135 **True.** Type 1: 80 per cent, inhibited young men, not physically dangerous. Type 2: 20 per cent, less inhibited, more psychopathic.

136 **False.** Castration has never been used in the UK.

137 **True.** It is used in hypersexuality, indecent exposure, unwanted fantasies, aggression, male contraception.

138 **True.**

139 **False.** They are the commonest.

140 **False.** The figure is 63 per cent.
Brooke D et al. *BMJ* 1996, **313**:1524–7.

141 **True.**

142 **True.** Testamentary capacity: must be of sound mind and (1) know the nature and extent of their property; (2) know the persons having a claim on it and the relative strengths of their claims; (3) be able to express themselves clearly and without ambiguity.

143 **True.** Bipolar II disorder: episodes of hypomania alternating with depression.

144 **True.** These risks are related to prolonged hypercortisolaemia.
Michelson D et al. *N Engl J Med* 1996; **335**(16):1176–81.

145 **True.**
Why Mothers Die. Report on Confidential Enquiries into Maternal Deaths in the United Kingdom 1997–1999. London, Royal College of Obstetricians and Gynaecologists, 2001.

146 **False.** The average duration is 7 months. It can persist for 1 year in 30 per cent of cases.

147 **False.** A washout period of 5 weeks is required.

148 **True.** But more adolescent boys are affected than adolescent girls.

149 **True.**

150 **False.** It is rarely associated with meaningful changes.
Stone J, Moran G. *Psychiatr Bull* 2003, **27**:171–2.

151 **True.** But it has not consistently been shown to have a significant role.

152 **False.** Cognitive decline in the course of schizophrenia is distinct from the neuropathological processes linked to β amyloid in Alzheimer's disease.

153 **True.** It can lead to subacute sclerosing pan-encephalitis.

154 **True.** The receptors are on the GABA channel.

155 **False.** Coprolalia (obscenities) is a feature. Coprophagia = ingesting faeces.

156 **False.** Currently there is no evidence base for benefits associated with this antioxidant.
Sauer J et al. *Cochrane Review*, Cochrane Database, 2004.

157 **True.** Plus few if any activities giving pleasure, emotional coldness, limited capacity to express warm feelings, no close or confiding relationships, little interest in sex, apparent indifference to praise/criticism, solitary, insensitivity to prevailing social norms and conventions.

158 **True.** It has been described in neuropathological and neuroimaging studies.

159 **True.**

160 **True.**
Harrington R et al. *Arch Gen Psychiatry* 1990, **47**(5):465–73.

161 **True.**
Mandy S et al. *Br J Psychiatry* 2001, **178**(Suppl. 40):S60–8.

162 **True.** Rett's syndrome is a pervasive developmental disorder affecting girls.
Amir RE et al. *Nat Genet* 1999, **23**:185–8.

163 **False.** 82 per cent are white, 7 per cent are black, 6 per cent are of mixed race, 2 per cent are Asian, 3 per cent are of 'other' races.
Department of Health. *Children Looked After in England 2000/2001.* London, HMSO, 2001.

164 **True.**
Ballenger JC et al. *J Clin Psychiatry* 1998, **59**:54–60.

165 **False.**

EMIs: Answers

1

1   A

2   E   Related to meta-analysis.

3   H   Receiver operating curve analysis can be used to determine the optimal cut-off point of a particular test.

2

1   A

2   B

3   D   A confounder is associated with both a determinant and the outcome but is not causative. An example would be the association between alcohol consumption and lung carcinoma. Smoking is the cause, alcohol is the confounder, commonly associated but not causative.

Note. Publication bias occurs with the publication of positive study findings over negative ones.

3

1   D

2   C

3   G   Whereas retrospective studies refer to outcomes occurring in the time before the study.

4   H   Parallel: e.g. two groups, (1) the study group, (2) control. Cross-over, more than two groups: subjects change groups during the study, e.g. from control to experimental group.

4

1   **D**   It usually occurs following alcohol withdrawal in dependent people. A short-lived confusional state is accompanied by a number of features including delusions, hallucinations, confusion, clouding of consciousness, agitation and fear.

2   **H**

3   **E**   Astereognosia is the inability to recognise an object through touch alone.

5

1   **F**   Brushfield's spots on the iris.

2   **D**

3   **E**   Also called fragile X syndrome.

4   **I**   More commonly called Prader–Willi syndrome.

5   **H**

6   **C**

6

1   **A**   It is of particular importance in Alzheimer's disease, which is associated with a cholinergic deficit (cholinergic hypothesis of memory dysfunction).

2   **B**

3   **C**

4   **D**   It is an anti-epileptic used increasingly for its mood-stabilising properties in mood disorders. It is (rarely) associated with serious skin reactions, including Stevens–Johnson's rash and also leucopenia.

PAPER 5

INDIVIDUAL STATEMENTS

1   Hypericum perforatum is superior to placebo in treating mild to moderate depression.   T | F

2   The pedunculopontine nuclei are relatively preserved in late-onset dementia.   T | F

3   Tetrahydrocannabinol has been shown to be effective and safe in the treatment of tics.   T | F

4   Reboxetine causes significant sexual dysfunction compared to the selective serotonin re-uptake inhibitors.   T | F

5   Negative predictive value is the proportion of people who are negative on the test who actually do not have the disorder.   T | F

6   Reliability is the ability of a test to measure what it purports to measure.   T | F

7   Saltatory nerve conduction is an abnormal finding in neurophysiological testing.   T | F

8   The resting cell membrane is relatively impermeable to Na$^+$.   T | F

9   There are five layers to the cerebral neocortex.   T | F

10   Late-onset obsessive–compulsive disorder in males has a poor prognosis.   T | F

11   Childhood disintegrative disorder is a rare form of disruptive behaviour associated with the pervasive developmental disorders.   T | F

12   The rate of suicide in the prison population is thought to be 10–20 times greater than that of the general population.   T | F

13   Post-ictal psychosis often begins after a lucid period.   T | F

14   Brief dynamic psychotherapy is rarely prescribed as first-line therapy.   T | F

15   The neuropsychological effects of stimulant abuse appear to persist for up to a year after having stopped usage.   T | F

16   Benzodiazepines increase beta waves.   T | F

17   Axis II in DSM-IV relates to general medical conditions.   T | F

18   Olanzapine is more likely to cause impotence than chlorpromazine.   T | F

19   Pseudobulbar palsy is not associated with the development of dementia.   T | F

20   Erectile dysfunction affects at least 2 per cent of men.   T | F

21   DSM-IV includes diagnostic criteria for male erectile disorder.   T | F

22   Dyspareunia has been reported to occur in up to 5 per cent of women aged 20–40 years.   T | F

23   The treatment of dyspareunia is usually pharmacological.   T | F

24   The reticular formation is associated with sleep and arousal.   T | F

25   Gilles de la Tourette syndrome is associated with coprolalia.   T | F

26   In an action potential, sodium and chloride ions pass into the cell and phosphate and potassium pass out.   T | F

27   High systolic blood pressure is a possible risk factor for Alzheimer's dementia.   T | F

28   Raised plasma homocysteine is an independent risk factor for the development of dementia and Alzheimer's disease.   T | F

29   Buprenorphine has little benefit in the treatment of heroin addiction.   T | F

30   Twilight states may be associated with unexpected violent acts.   T | F

31   Sleep walking and night terrors occur in REM sleep.   T | F

32   Paradoxical sleep is associated with abnormal EEG rhythms.   T | F

33   The hypothalamic nuclei can be divided into anterior, intermediate and posterior nuclei.   T | F

34   Stimulation of the septohippocampal formation causes sedation.   T | F

35   Treatment with low-dose risperidone reduces aggression, agitation and psychosis in dementia.   T | F

36   The substantia nigra is a structure within the midbrain.   T | F

37   The facial nerve carries parasympathetic neurones.   T | F

38 In manic defence the individual defends himself against anxiety,   T | F
    isolation and psychosis.

39 Hyponatraemia is associated with selective serotonin re-uptake   T | F
    inhibitor and venlafaxine use in the elderly.

40 Changes in thalamus blood flow have been shown following   T | F
    treatment response to venlafaxine.

41 Globus hystericus is a fear that the world is 'going mad'.   T | F

42 Family therapy for schizophrenia has been associated with   T | F
    reduced relapse rates.

43 Recurrent major depression is a risk factor for subclinical   T | F
    atherosclerosis in middle-aged women.

44 Measuring cerebrospinal fluid tau protein can help differentiate   T | F
    old-age depression from Alzheimer's disease.

45 The anterior spinal artery is a branch of the basilar artery.   T | F

46 Wilson's disease is a cause of reversible dementia.   T | F

47 Lip smacking and chewing are seen in temporal lobe epilepsy.   T | F

48 There is diffuse flattening of the EEG in Huntington's disease.   T | F

49 Reduction in grey matter is a rare finding in schizophrenia.   T | F

50 Frontal lobe function involves an ability to switch mental tasks.   T | F

51 Half of narcolepsy sufferers have major affective disorders.   T | F

52 Agoraphobia follows a fluctuating course.   T | F

53 A conversion reaction often relieves anxiety in conversion disorders.   T | F

54 Phantom limb pain is a long-standing conversion disorder.   T | F

55 Weight gain has been reported with flupentixol.   T | F

56 Object agnosia is associated with a lesion of the left occipital   T | F
    cerebral cortex.

57 The incidence of schizophrenia is between 18 and 50 per 100 000   T | F
    of the population.

58 The Wilcoxon test is suitable for detecting differences between   T | F
    two repeated measures.

59 In older hospitalised patients, depression seems to be associated with a greater occurrence of adverse drug reactions.   T   F

60 Steroids may improve negative, depressive and anxiety symptoms in schizophrenia.   T   F

61 Dementia is inevitable in all those living beyond 90 years of age.   T   F

62 Eighty per cent of depressive episodes receive no treatment.   T   F

63 Dreaming only occurs in REM sleep.   T   F

64 Accidental injuries are common in people with somnambulism.   T   F

65 Tangles are extracellular bodies found in Alzheimer's dementia.   T   F

66 Lithium undergoes mainly renal elimination.   T   F

67 Benzodiazepines increase fast-wave EEG activity.   T   F

68 Learning-disabled people with schizophrenia often respond poorly to neuroleptic medication.   T   F

69 Gilles de la Tourette syndrome is associated with specific EEG abnormalities.   T   F

70 Pimozide is used commonly for the treatment of simple tics in children.   T   F

71 Mortality following ECT is similar to that following general anaesthesia for small procedures.   T   F

72 The majority of episodes of delirium recover in less than 4 weeks.   T   F

73 A diagnosis of delirium is unlikely if there is disturbance of the sleep/wake cycle.   T   F

74 Nystagmus is a recognised feature of acute alcohol intoxication.   T   F

75 'Forced thinking' can be a symptom in temporal lobe epilepsy.   T   F

76 Increased appetite can develop within a few hours of cocaine withdrawal.   T   F

77 Lithium inhibits tau phosphorylation.   T   F

78 There is an association between dental caries and antidepressants.   T   F

79 Apolipoprotein D has been implicated in Alzheimer's disease among African Americans.   T   F

80 The therapist indulges in Socratic questioning in cognitive therapy.　　　T | F

81 Faecal soiling in an infant is associated with faecal retention.　　　T | F

82 There is an association between sexual pleasure and pathological fire setting.　　　T | F

83 The prevalence of dyslexia is higher in inner city areas.　　　T | F

84 Periods of REM sleep lengthen during the natural course of sleep.　　　T | F

85 Sexual exhibitionism usually progresses to sexual attacks.　　　T | F

86 Elderly people with subjective memory complaints and objective cognitive impairment have a high risk of developing Alzheimer's dementia.　　　T | F

87 A three-step procedure (self-report memory complaints, tests of global cognitive functioning, and domain-specific cognitive tests) has a positive predictivity of 90 per cent for Alzheimer's disease and dementia at 3 years.　　　T | F

88 A traumatic brain injury during childhood or adolescence could more than double an individual's chance of developing psychiatric disorders.　　　T | F

89 Impaired serotonergic function does not contribute to cognitive decline in Alzheimer's disease.　　　T | F

90 Approximately 10 per cent of reversible cases of dementia are truly reversible.　　　T | F

91 Fluctuating cognition, visual hallucinations and Parkinsonism are all required for a diagnosis of dementia with Lewy bodies.　　　T | F

92 Profound cerebellar signs are found in Creutzfeldt–Jakob disease.　　　T | F

93 NMDA receptor antagonists are used in the treatment of dementia.　　　T | F

94 Autistic children lack 'theory of mind'.　　　T | F

95 Case control studies are seldom subject to recall bias.　　　T | F

96 Cocaine damages brain circuits associated with the sense of pleasure.　　　T | F

97   Eighty per cent of manic patients respond to lithium treatment.   T | F

98   Cerebellar degeneration occurs in association with chronic alcohol abuse.   T | F

99   Third ventricle petechiae occur with chronic alcohol consumption.   T | F

100   In Alzheimer's disease, low levels of acetyltransferase are found in the nucleus basalis of Meynert.   T | F

101   There is an association between anorexia nervosa and shoplifting.   T | F

102   Psychotropics can play a role in the pathologically jealous individual.   T | F

103   Venlafaxine is associated with polyarthralgia.   T | F

104   White matter hyperintensities observed on MRI scanning are common in late-onset depression.   T | F

105   Venlafaxine can cause a dose-related rise in blood pressure.   T | F

106   The Brief Psychiatric Rating Scale (BPRS) is unsuitable for rating minor psychiatric disorders.   T | F

107   Conner's rating scale is used as a measure of hyperactivity.   T | F

108   An REM latency of 90 minutes is found in normal adults.   T | F

109   Children of school-going age are the commonest subjects of Munchausen's syndrome by proxy.   T | F

110   Fluctuations in attention would indicate a diagnosis of Alzheimer's disease rather than dementia with Lewy bodies.   T | F

111   A clinically significant interaction is likely between a tricyclic antidepressant and tranylcypromine.   T | F

112   Transcranial magnetic stimulation can reduce auditory hallucinations.   T | F

113   Hallucinations are more frequent in dementia with Lewy bodies than in Alzheimer's disease.   T | F

114   In autism, failure to develop useful speech by the age of 5 years is a predictor of poor outcome.   T | F

115   Approximately 5 per cent of 3 year olds have a specific language delay.   T | F

116   Methylphenidate inhibits the dopamine transporter mechanism.   T | F

117   Tics affect 20 per cent of children.   T | F

118   Tics are more pronounced in sleep.   T | F

119   Tricyclic antidepressants are useful in childhood depression.   T | F

120   Prolonged interferon-alpha therapy for hepatitis C may induce depressive symptoms and major depression.   T | F

121   Childhood-onset schizophrenia is associated with progressive loss of cerebellar volume in adolescence.   T | F

122   Rivastigmine is a selective inhibitor of buteryl cholinesterase.   T | F

123   Elderly women are more likely than their male counterparts to commit suicide.   T | F

124   Late-onset paraphrenia has an equal sex distribution.   T | F

125   The prevalence of Alzheimer's disease is greatest in rural areas.   T | F

126   Diogenes syndrome is associated with end-stage personality disorder.   T | F

127   Leukoariosis is associated with cognitive decline.   T | F

128   Parietal lobe symptoms are a poor prognostic indicator in Alzheimer's disease.   T | F

129   Formication can occur in cocaine usage.   T | F

130   Alcohol misuse leads to demyelination of the corpus callosum.   T | F

131   Very late-onset schizophrenia-like psychosis is diagnosed at age 80 or above.   T | F

132   In the elderly, drugs have shorter half-lives.   T | F

133   Liquid Ecstasy is an illicit drug used by 'ravers and body builders'.   T | F

134   Bilateral hippocampal destruction leads to Korsakoff's syndrome.   T | F

135   Peripheral neuropathy is a clinical feature of Wernicke's encephalopathy in less than 30 per cent of cases.   T | F

136   The falsification of memory in clear consciousness occurs in Wernicke's encephalopathy.   T | F

137   Omega-3 fatty acids may have a role in treating borderline   T | F
      personality.

138   Caution is needed before prescribing an acetylcholinesterase   T | F
      inhibitor to a patient with atrial fibrillation.

139   Charles–Bonnet syndrome is characterised by poorly formed   T | F
      visual hallucinations.

140   Depressed elderly people do not benefit from a team approach to T | F
      care.

141   Nalbuphine is used by athletes to control anxiety before a   T | F
      competition.

142   Finkelhor created a risk factor checklist for sexual abuse.   T | F

143   In forensic samples, borderline personality disorder shows   T | F
      comorbidity with antisocial personality disorder.

144   Dextromoramide is an opioid drug.   T | F

145   Tetrahydrocannabinol and its metabolites can be detected in the T | F
      urine for several weeks after use.

146   Cocaine is a central nervous system depressant inducing calmness T | F
      and sleep.

147   Post-stimulant perception disorder (or 'flashbacks') frequently T | F
      occurs with cocaine usage.

148   Phencyclidine is usually smoked, but can be taken orally,   T | F
      intravenously or inhaled.

149   Lamotrigine may be an effective option for the treatment of   T | F
      depression in epilepsy.

150   Divalproex is used in the treatment of acute mania associated   T | F
      with bipolar disorder.

151   Men who experience sexual dysfunction due to the use of   T | F
      selective or non-selective serotonin re-uptake inhibitors may
      benefit from sildenafil.

152   Incremental validity indicates whether the measurement being   T | F
      assessed is superior to other measurements in approaching true
      validity.

153   Face validity is a subjective judgement as to whether a test or T | F
      measure appears to measure the feature in question.

154 The bias occurring when the researcher has clues about whether the subject is in the case or control group is called selection bias. T | F

155 Studies are generally considered worthwhile if the power is at least 80 per cent. T | F

156 The relative risk is the ratio of the risk of an outcome in one experimental group divided by the risk of the outcome in the other group. T | F

157 The Z statistic represents the deviation from the mean value in standard deviation units. T | F

158 The process of ensuring that all subjects recruited into a trial have equal chances of being allocated to the treatment or control groups is called distribution. T | F

159 Departures from the symmetry in a frequency distribution are known as skewness. T | F

160 In a normal distribution, mode > median > mean. T | F

161 The kappa is a chance-corrected measure of agreement. T | F

162 In an intention to treat analysis, data only on those who actually received the treatment to which they were allocated are included. T | F

163 Likelihood ratio is equal to the sensitivity over 1 minus the specificity. T | F

164 Bias is a random error. T | F

165 Kurtosis is a normal distribution that appears too flat or too 'peaky'. T | F

EMIs

1

A	Stein–Leventhal disease	G	Kennedy's syndrome
B	Parkinson's disease	H	Dementia with Lewy bodies
C	Guillain–Barré syndrome	I	Dysmnesic syndrome
D	Multiple sclerosis	J	Neurosyphilis
E	Pseudobulbar palsy	K	None of the above
F	Wilson's disease		

Choose the most suitable of the above for each of the following.

1 A patient with this disorder is found to have a trinucleotide repeat.

2 A patient with this condition is told that he has associated Uthoff's phenomenon.

3 A patient's decline is arrested with penicillin, with possible slight improvement.

4 This condition is associated with periungual fibromas.

2

A	Trichotillomania	G	School refusal
B	Affective disorders	H	Asperger's syndrome
C	Hyperkinetic disorder	I	Pervasive developmental
D	Conduct disorder		disorders
E	Autism	J	Early-onset schizophrenia
F	Tic disorder		

Which of the above disorders applies to each of the following?

1   This is the commonest childhood psychiatric disorder.

2   A subtype of this disorder is called oppositional defiant disorder, according to ICD-10.

3   Strict exclusion diets have been used in an attempt to treat the symptoms of this condition, with varying success.

4   The onset of this condition is associated with three age-related peaks in incidence.

5   Its severity gradually improves so that by early adulthood there is commonly complete or partial resolution.

3

A   Ash leaf-shaped macules which
    fluoresce under ultraviolet light
B   Bilateral acoustic neuromas
C   Maculopapular rash
D   Blue naevi
E   Cherry-red spot
F   Axillary freckling

G   Hypertension
H   Ejection systolic murmur
I   Transposition of the great
    arteries
J   Broad tongue
K   Brushfield's spots

Answer the following.

1   A 7 year old with mental retardation, seizures and facial angiomas is found
    to have what else on physical examination?

2   A 12 year old is found to have more than six café au lait macules and four
    Lisch nodules. In addition, she was found to have what on physical
    examination?

3   A 22 year old found to have an abnormality on chromosome 22 presents
    with a cranial meningioma and spinal tumour in addition to what from the
    above?

4

A	7–10 days	F	15–25 days
B	1–2 days	G	25–35 days
C	3–9 days	H	40–50 days
D	4–7 days	I	50–65 days
E	10–15 days		

Match each of the following depot preparations with its time to peak plasma concentration following administration.

1   Flupentixol decanoate

2   Fluphenazine decanoate

3   Haloperidol decanoate

4   Pipothiazine palmitate

5   Zuclopenthixol decanoate

6   Zuclopenthixol acetate

5

A	Cannabis	F	Temazepam
B	Cocaine	G	Methadone
C	Diazepam	H	Amphetamine
D	Heroin	I	Ecstasy
E	LSD	J	Codeine

Match the appropriate drug to each of the statements below.

1   This drug has an elimination half-life of 28 hours.

2   This drug has an elimination half-life of 48 hours.

3   This drug has an elimination half-life of 2 minutes.

4   This drug has an elimination half-life of 6 hours.

5   This drug has an elimination half-life of 1 hour.

6   This drug has an elimination half-life of 10 hours.

6

A	Feighner criteria	G	Kendall's correlation coefficient
B	Kappa coefficient		
C	Friedman test	H	Parson's correlation coefficient
D	Kruskal–Wallis test	I	Ratio scaling
E	Mann–Whitney test	J	Discriminant validity
F	Wilcoxon test		

Which of the above is:

1   A non-parametric test of differences in repeated observations on more than two occasions?

2   A non-parametric test of differences in medians of two or more independent groups?

3   A non-parametric correlation coefficient?

4   A non-parametric test of the difference between the medians of two paired sets of observations (usually before and after an intervention)?

INDIVIDUAL STATEMENTS: Answers

1   **True.** St John's wort.

2   **False.** The pedunculopontine nuclei are a part of the cholinergic system, projecting to the thalamus. They commonly degenerate.

3   **True.**
    Muller-Vahl KR et al. *J Clin Psychiatry* 2003, **64**(4):459–65.

4   **False.** It might be of benefit for patients at risk of sexual dysfunction with selective serotonin re-uptake inhibitors.
    Clayton AH et al. *Int Clin Psychopharmacol* 2003, **18**(3):151–6.

5   **True.**

6   **False.** This is the definition of validity. Reliability is the level of agreement between different sets of observations.

7   **False.** It is normal nerve conduction, jumping from nodes of Ranvier.

8   **True.**

9   **False.** There are six layers: molecular/plexiform, external granular, external pyramidal, internal granular, internal pyramidal/ganglionic, multiform/polymorphic.

10 **False.**

11 **False.** Childhood disintegrative disorder is a form of dementia occurring in early life, often associated with lipidoses or other progressive brain pathology.

12 **True.**

13 **True.** The commonest psychotic features in such patients are hallucinations and delusions.

14 **False.**

15 **True.**
    Toomey R et al. *Arch Gen Psychiatry* 2003, **60**:303–10.

16 **True.** Barbiturates and benzodiazepines increase beta and reduce alpha activity.

17   **False.** Axis II relates to personality.

18   **False.**

19   **False.** Pseudobulbar palsy involves spastic weakness of pharyngeal musculature causing dysphasia and dysarthria with emotional lability and dementia.

20   **True.**

21   **True.**

22   **False.** Dyspareunia occurs in 10–20 per cent of 20–40 year olds. It is genital pain associated with sexual intercourse.

23   **False.** It is usually psychological.

24   **True.** Brainstem formation is probably involved in sleep, consciousness, movement and motivation.

25   **True.** Coprolalia is verbalised obscenities.

26   **False.** Sodium and phosphate move in, potassium and chloride pass out of the cell.

27   **True.** In a cohort of 1270 elderly people, 339 developed dementia over a 6-year period. Both low diastolic and high systolic pressures were associated with increased risk of Alzheimer's disease and dementia. Qui C et al. *Arch Neurol* 2003, **60**:223–8.

28   **True.**
      Seshadri S et al. *N Engl J Med* 2002, **346**:476–83.

29   **False.** Buprenorphine has both agonist and antagonist activity and may be effective in the treatment of heroin addiction. It can also precipitate withdrawal symptoms in those with opioid dependency. Kakko J et al. *Lancet* 2003, **361**:662–8.

30   **True.** They are associated with temporal lobe epilepsy and organic states.

31   **False.** They occur in non-REM sleep. REM sleep is associated with nightmares.

32   **False.** Paradoxical sleep is fast mixed frequency activity of low voltage (similar to awake trace) occurring in REM sleep – hence its name. It is a normal finding.

33   **True.**

34   **False.** The septohippocampal formation is a 'chief pleasure pathway'. It is also a cholinergic supply to the hippocampus.

35   **True.** These are the findings from a trial of 345 patients randomised to either risperidone or placebo. However, recent guidance suggests that the increased risk of cerebrovascular accidents associated with certain atypical neuroleptics means that they are now contraindicated for behavioural management in patients with dementia!
     Brodaty H et al. *J Clin Psychiatry* 2003, **64**:134–43.

36   **True.** The midbrain comprises the tectum, tegmentum (cerebral peduncle, red nucleus and substantia nigra) and the CNS cerebri.

37   **True.**

38   **False.** Manic defence is a form of defensive behaviour against anxiety, guilt and depression by (1) denial, (2) phantasy of omnipotent control, (3) identification, (4) projection.

39   **True.** In a study of 199 elderly patients taking either venlafaxine or a selective serotonin re-uptake inhibitor, 39 per cent were hyponatraemic. These medications can cause hponatraemia in any individuals.
     Kirby D et al. *Int J Geriatr Psychiatry* 2002, **17**:231–7.

40   **True.**
     Davies J et al. *Am J Psychiatry* 2003, **160**:374–6.

41   **False.** It is subjective difficulty in swallowing.

42   **True.**

43   **True.** This was shown in a study of 336 women in the USA.
     Jones DJ et al. *Arch Gen Psychiatry* 2003, **60**:153–60.

44   **True.**
     Buerger K et al. *Am J Psychiatry* 2003, **160**:376–9.

45   **False.** The anterior spinal and vertebral arteries join to form the basilar artery. The basilar artery has five branches (superior cerebellar, pontine, basilar, labyrinthine and anterior inferior cerebellar).

46   **True.** The dementia is potentially reversible, although some cognitive deficit is common.

47 **True.** Other features are illusions, hallucinations, déjà vu, jamais vu, depersonalisation, and affective experiences.

48 **True.**

49 **False.**

50 **True.** In the Stroop test, the word 'blue' is written in red on paper and the subject is asked to say what colour it is.

51 **True.** Narcolepsy: excessive daytime somnolence, sudden onset of REM sleep accompanied by cataplexy or sleep paralysis. Twenty-five per cent of sufferers have associated hypnopompic or hypnogogic hallucinations.

52 **True.**

53 **True.**

54 **False.** It is the sensation of pain in a body part that has been lost or amputated. Sensation usually fades with time.

55 **True.**

56 **True.** Left occipital cortex: right homonymous hemianopia, alexia, colour naming defect, object agnosia. Right occipital cortex: left homonymous hemianopia, visual illusions and hallucinations, loss of topographical memory and visual orientation.

57 **True.**

58 **True.**

59 **True.**
Onder G et al. *Arch Intern Med* 2003, **163**:301–5.

60 **True.** Dehydroepiandrosterone (DHEA) was demonstrated to improve these symptoms in a small study of 30 patients.
Strous RD et al. *Arch Gen Psychiatry* 2003, **60**:133–41.

61 **False.** Of those people who live beyond 90 years of age, many are unaffected by dementia.
Boeve B et al. *Neurology* 2003, **60**:477–80.

62 **True.**

63 **False.** Dreaming usually occurs in REM, but can occur in non-REM sleep.

64 **False.** Somnambulists (sleepwalkers) occasionally harm themselves, so their environment should be made safe before they sleep.

65 **False.** Tangles are intracellular. Remember people usually 'tango' inside.

66 **True.** Clearance is therefore reduced in the elderly and with renal impairment.

67 **True.** They increase β and reduce α. In overdose there is prominent fast activity.

68 **True.**

69 **False.** It is associated with non-specific EEG changes.

70 **False.** It is not commonly used in the treatment of tics. It has been used in the treatment of Tourette's syndrome, but is not recommended for use in children and is associated with ECG abnormalities.

71 **True.**

72 **True.**

73 **False.** This is a characteristic finding, as are daytime drowsiness, nocturnal worsening of symptoms, disturbing dreams or nightmares.

74 **True.**

75 **True.** It is a compulsion to think on specific topics.

76 **True.** Others include dysphoric mood, fatigue, vivid unpleasant dreams, insomnia/hypersomnia and psychomotor retardation or agitation (DSM-IV).

77 **True.** In cell lines, lithium has been shown to inhibit tau phosphorylation. It is thought that inhibition of glycogen synthase kinase-3 accounts for this effect. Lithium is now being investigated for a potential therapeutic role in Alzheimer's dementia.

78 **True.** Antidepressants can cause dry mouth (xerostomia). This can lead to oral complications, including dental caries.
Keen JJ Jr. *J Am Dent Assoc* 2003, 134:71–80.

79   **True.** Apolipoprotein D has recently been implicated in Alzheimer's disease among African-American people. Further studies will be needed to confirm this.

80   **True.**

81   **True.**

82   **True.**

83   **True.**

84   **True.**

85   **False.** Most exhibitionists do not commit violent sexual acts, nor do they interfere with children.
      Rooth G. *Arch Sex Behav* 1973, 2(4):351–63.

86   **True.** Alzheimer's disease is thought to affect 50–60 per cent of all dementia patients.

87   **True.** In a study of 1435 people aged 75–95, single questions were asked about memory complaints and assessment was by MMSE and neuropsychological testing.
      Palmer K et al. *BMJ* 2003, **326**:245.

88   **True.**
      Timonen N et al. *Psychiatry Res* 2002, 113:217–26.

89   **False.** The serotonergic system is impaired in Alzheimer's disease, which may contribute to both cognitive and non-cognitive symptoms.
      Porter RJ et al. *Psychol Med* 2003, 33:41–9.

90   **True.** By the time the majority are diagnosed, a degree of irreversible cognitive impairment will have occurred.

91   **False.** Two of these are required for probable Lewy body dementia, one for a possible diagnosis.

92   **True.** In addition to extrapyramidal signs and myoclonus.

93   **True.** Memantine is used in the treatment of Alzheimer's disease.

94   **True.** This is the ability to form an idea of what others are thinking.

95 **False.**

96 **True.** Dopamine neurones in cocaine users may be damaged or destroyed. This may predispose to depression.
Littel KY et al. *Am J Psychiatry* 2003, **160**:47–55.

97 **True.** But response can take up to 3 weeks. Other medications are often needed acutely for more immediate response (neuroleptics, benzodiazepines).

98 **True.**

99 **True.**

100 **True.** It relates to the cholinergic deficit in Alzheimer's dementia.

101 **True.** Sufferers are most likely to steal food.

102 **True.** Especially if they are of a delusional nature. Response to treatment is often disappointing and drug adherence poor.

103 **True.** It is also associated with myalgia.

104 **True.**

105 **True.** Blood pressure monitoring is advisable, especially if there is pre-existing hypertension.

106 **True.** It is an instrument with 16 items scored on a seven-point scale.

107 **True.**

108 **True.**

109 **False.** It usually involves younger children.

110 **False.** This is more likely to be dementia with Lewy bodies.

111 **True.** Co-administration of a tricyclic antidepressant and an MAOI can lead to a toxic reaction.

112 **True.** Repetitive transcranial magnetic stimulation reduced auditory hallucinations resistant to conventional treatment in a small study of schizophrenic and schizoaffective patients.
Hoffman RE et al. *Arch Gen Psychiatry* 2003, **60**:49–56.

113 **True.**

114 **True.** Those with an IQ of 70 or above and who use communicative language by the age of 5 years have a better prognosis.

115 **True.**

116 **True.** It is a central nervous system stimulant used for attention deficit hyperactivity disorder. Dexamphetamine is also used.

117 **False.** They affect 5–10 per cent of children.

118 **False.** They disappear in sleep.

119 **False.** They fell out of general favour after meta-analyses showed they were no better than placebo in depression. They are also associated with significant side effects.

120 **True.**
Bonaccorso S et al. *J Affect Disord* 2002, **72**:237–41.

121 **True.** In 50 childhood-onset cases and 50 controls, affected individuals showed greater loss of cerebellar volume.

122 **False.** It is a dual inhibitor of acetylcholinesterase and buterylcholinesterase.

123 **False.**

124 **False.** It is 4–20 times more common in women (depending on which study).

125 **False.** Alzheimer's has been shown by a number of studies to be commoner in urban areas.
Baker FM et al. *Int J Geriatr Psychiatry* 1993, 8:379–85.

126 **True.** Diogenes syndrome has been considered by some to be the end-stage of a personality disorder, manifesting itself in the form of senile reclusiveness.

127 **True.** It is a form of vascular dementia associated with reduced white matter density.

128 **True.** Others include marked language impairment and poor cognitive function.

129 **True.** Formication is tactile hallucinations.

130 **True.** Clinical features of the Marchiafava-Bignami disease include ataxia, epilepsy, dysarthria and impaired consciousness. There is demyelination of the corpus callosum, optic tract and cerebellar peduncles.

131 **False.** Criteria for very late onset schizophrenia:
> 60 years
fantastic, persecutory, referential or grandiose delusions
+/- hallucinations
absence of primary affective disorder
MMSE no less than 25/30
no clouding of consciousness
no neurological illness or alcohol dependence.

132 **False.** They have longer half-lives as a function of reduced renal clearance and distribution.

133 **True.** Also called gamma-hydroxybutyrate, it is an endogenous fatty acid found in all cells; it potentiates cerebral dopaminergic systems.

134 **True.**

135 **False.** Peripheral neuropathy occurs in up to 80 per cent of cases.

136 **False.** This is confabulation and is a feature of Korsakoff's syndrome.

137 **True.** According to the results of a small study suggesting that E-EPA may be efficacious for the treatment of moderately disturbed women with borderline personality disorder.
Zanarini MC, Frankenburg FR. *Am J Psychiatry* 2003, **160**:167-9.

138 **True.** Caution is needed in patients with sick sinus syndrome or conduction abnormalities, but it is not absolutely contraindicated.

139 **False.** It occurs in the elderly. Defining characteristics: well-formed visual hallucinations (vivid), preserved intellectual competence, clear consciousness, insight maintained.

140 **False.** Forty-five per cent of people in the team approach group had at least a 50 per cent reduction in symptoms of depression, compared with only 19 per cent in the standard care group.
Unutzer J et al. *JAMA* 2002, **288**:2836-45.

141 **True.** Nalbuphine hydrochloride (Nubain) is an opioid used to increase the pain threshold or as an anti-anxiety drug. It can cause both dependence and psychiatric side effects.

142 **True.** The eight strongest independent predictors are:
having a stepfather
having lived without mother
not being close to mother
mother not finishing secondary school
having a sex-punitive mother
no physical affection from father
low income family
two or fewer friends in childhood.

143 **True.**
Tyrer P et al. *Br J Psychiatry* 2003, 182(Suppl. 44):s1–35.

144 **True.**

145 **True.**

146 **False.** Cocaine is a central nervous system stimulant.

147 **False.** Post-hallucinogen perception disorder ('flashbacks') occurs with hallucinogenic drugs such as LSD and DMT (dimethyltryptamine).

148 **True.** Phencyclidine (PCP) is a hallucinogen with both depressant and stimulant effects.

149 **True.** A small study concluded that lamotrigine may be a good option for patients with epilepsy and concomitant mood disorders.
Kalogjera-Sackellares D, Sackellares JC. *Epilepsy Behav* 2002, 3:510–16.

150 **True.** Divalproex sodium is a compound of sodium valproate and valproic acid.

151 **True.**
Numberg HG et al. *JAMA* 2003, 289:56–64.

152 **True.**

153 **True.**

154 **False.** This is observer bias.

155 **True.** Power is the probability of demonstrating a significant difference between groups when one exists.

156 **True.**

157 **True.** The Z statistic is the non-standardised normal deviate.

158 **False.** This is randomisation, the point being to eliminate the bias of influencing subject allocation.

159 **True.** A frequency distribution may be either positively or negatively skewed.

160 **False.** In a normal distribution, mode = median = mean.

161 **True.** It is primarily a measure of reliability.

162 **False.** Data on all randomised subjects are analysed within the groups to which they have been assigned whether or not they actually received the treatment to which they were allocated.

163 **True.** The likelihood ratio is the likelihood that a positive test result will be observed in a patient with, as opposed to a patient without, the disorder.

164 **False.** Bias is an example of a systematic error.

165 **True.**

EMIs: Answers

1

1   **G**   An inherited neuromuscular disorder with increased CAG repeats and associated with psychiatric disturbances.

2   **D**   Symptoms are exacerbated by heat.

3   **J**

4   **K**   This is tuberous sclerosis.

2

1   **D**   Four per cent prevalence in the Isle of Wight study; boys:girls 3:1.

2   **D**

3   **C**

4   **G**   5–6 years (starting school), 11–12 (secondary school) and early teens.

5   **F**   Up to 90 per cent improve within 5 years.

3

1   **A**   This is tuberous sclerosis, affecting 1 in 6000 births. Skin lesions include ash-leaf macules, fibrous forehead plaques, adenoma sebaceum and periungual fibromata.

2   **F**   This is characteristic of type I neurofibromatosis, a neurocutaneous syndrome (Von Recklinghausen). Complications include malignant skin lesions, spinal tumours, gastrointestinal neurofibromas and epilepsy.

3   **B**   This is type II neurofibromatosis. It is rare in children, with minimal skin signs. It is associated with acoustic neuromas, cranial meningiomas and spinal tumours.

4

1   A

2   B

3   C

4   A

5   D

6   B

In clinical practice we consult the formulary, but an idea of peak plasma concentration is important in terms of clinical response and adverse effects.

5

1   A

2   C

3   D

4   I

5   B

6   F

6

1   C

2   D   Also called the Kruskal–Wallis ANOVA.

3   G   A correlation coefficient is a summarising value describing a relationship between variables.

4   F

PAPER 6

INDIVIDUAL STATEMENTS

1   An independent variable (in addition to the one under   T   F
    examination) that has a systematic influence on the dependent
    variable is called an attributer.

2   The null hypothesis is a prediction that there is no relationship   T   F
    between the independent and dependent variables.

3   A type I error is the error of rejecting the null hypothesis when it   T   F
    is true.

4   To reduce the chance of a type I error one chooses a lower value   T   F
    of p (probability).

5   χ^2 test is a significance test which has the effect of comparing   T   F
    two or more independent proportions.

6   The power of a study is determined by dividing the standard   T   F
    deviation by the clinically significant difference and then cross-
    referencing this ratio with a table of power for a given number of
    subjects.

7   The number needed to treat is $1 \times$ absolute risk reduction.   T   F

8   The positive predictive value of a test is the proportion of subjects   T   F
    with positive test results who actually have the disease.

9   Asymmetrical funnel plots indicate either publication bias or   T   F
    exaggeration of treatment effects in small studies of low
    quality.

10 Failure to reject a false null hypothesis is called a type II error.   T   F

11 A trial is 'double blind' if neither the subject nor the investigator   T   F
    knows the outcome.

12 Validity refers to the extent to which a measure really does   T   F
    measure what it sets out to measure.

13 Construct validity assesses the extent to which a new measure can   T   F
    predict future variables.

14 The variation of individual studies according to the participant   T   F
    characteristics, study design and conduct of the study is termed
    heterogeneity.

15 The odds ratio closely approximates to the risk ratio in a cohort   T | F
    study.

16 In epidemiological studies, four possible explanations for an   T | F
    association exist.

17 Cohort studies are prone to selection bias arising from incomplete   T | F
    follow-up and confounding.

18 Power is the probability that a type II error will not be made.   T | F

19 Inadequate sample size is indicated by the relatively large width   T | F
    of the corresponding confidence interval.

20 The p value on its own implies a great deal about the magnitude   T | F
    of any difference between treatments.

21 Psychostimulants have been used since the 1930s to attenuate   T | F
    hyperactivity and improve cognitive performance.

22 Kanner first described autism in a series of papers in the early   T | F
    nineteenth century.

23 Seventeen per cent of 5 year olds have functional (non-organic)   T | F
    enuresis.

24 An IQ range of 20–34 represents a profound learning disability.   T | F

25 A flattened occiput is characteristic of Down's syndrome.   T | F

26 In Down's syndrome, the feet have a wide gap between the first   T | F
    and second toes and a plantar furrow extending posteriorly.

27 Regular attempts to defecate plus laxatives are usually successful   T | F
    in the management of encopresis.

28 A crime requires both an actus rea and mens rea.   T | F

29 Not guilty of a criminal act by reason of insanity would require   T | F
    fulfilment of the McManus Rules (1815).

30 Infanticide is when a parent causes the death of their child under   T | F
    12 months of age.

31 Fitness to plead is dependent upon the fulfilment of the Pritchard   T | F
    Criteria.

32 Of those children diagnosed with autism, 20 per cent have   T | F
    medical conditions known to be associated with autism.

33   Rett's syndrome occurs in 90 per cent of girls.   T | F

34   MRI may have a role in the diagnosis of psychosis before the   T | F
     expression of frank psychotic symptoms.

35   Adults with attention deficit hyperactivity disorder can be   T | F
     differentiated from healthy individuals using EEG measures.

36   The five-axis classification of the International Classification of   T | F
     Diseases was developed to provide a thorough and complete
     description of a patient's psychiatric illness, medical problems and
     overall functioning.

37   Task-focused attention leads to more social anxiety in blushing   T | F
     anxious, socially anxious and social phobic individuals.

38   Inositol depletion has been implicated in the mechanism of action   T | F
     of lithium, carbamazepine and valproate.

39   Late-life depression has a greater negative outcome for men than   T | F
     for women.

40   Sildenafil citrate is used in male sexual dysfunction.   T | F

41   Cerebrospinal levels of tau protein and β-amyloid 42 are putative   T | F
     biomarkers to predict dementia outcome.

42   The majority of patients with multiple sclerosis have comorbid   T | F
     depression.

43   Nefazodone has been useful in pathological gambling.   T | F

44   Mothers under the age of 35 who have children with Down's   T | F
     syndrome are at increased risk of Alzheimer's disease.

45   Gait abnormalities are associated with an increased risk of   T | F
     vascular dementia.

46   Melatonin is produced by the pineal gland.   T | F

47   Children under 16 years of age do not have the capacity to   T | F
     consent to an intervention.

48   In Huntington's chorea, disease onset is usually in the 50s.   T | F

49   In Huntington's disease, there is bilateral wasting of caudate   T | F
     and putamen nuclei with generalised fronto-temporal
     wasting.

50 Lamotrigine-induced arrhythmias are minimised by gradual dose titration. T | F

51 Hypometabolism in the caudate nuclei has been demonstrated in women with somatisation disorder. T | F

52 Restlessness is a side effect of *Hypericum perforatum.* T | F

53 Lower levels of serotonin alpha-2 (5HT2a) receptors in the prefrontal cortex and hippocampus have been linked to suicidal behaviour. T | F

54 There is an association between neurological soft signs in childhood and adult psychosis. T | F

55 Low interleukin-6 levels are associated with major depression in cancer patients. T | F

56 An increase in acute psychiatric presentations was noted during the transition into the new millennium. T | F

57 The selective serotonin re-uptake inhibitors aggravate the motor symptoms of depressed people with Parkinson's disease. T | F

58 Outpatients with post-traumatic stress disorder (PTSD) and comorbid borderline personality disorder (BPD) have a more severe clinical profile than outpatients with PTSD or BPD alone. T | F

59 Donepezil reduces REM latency in depressed patients. T | F

60 Adolescent depression is more common among girls than boys. T | F

61 Transdermal selegiline shows benefits for major depression. T | F

62 Hydroxyzine is ineffective in generalised anxiety disorder. T | F

63 Lithium uncouples GTP-binding protein from stimulation of adenylate cyclase and/or phospholipase-c. T | F

64 Meta-analyses show that psychological treatments produce effect sizes of 0.8–1.0. T | F

65 Galantamine was originally derived from the tulip. T | F

66 Lesions in the floor of the fourth ventricle are found in Wernicke's encephalopathy. T | F

67 Supratentorial hypertrophy is seen in Korsakoff's state. T | F

68   Olanzapine is safe in the elderly patient with dementia.   T | F

69   Thioridazine increases the QT interval.   T | F

70   Amiodarone increases the QT interval.   T | F

71   Tricyclic antidepressants increase the QT interval.   T | F

72   Hypocalcaemia reduces the QT interval.   T | F

73   Quinidine increases the QT interval.   T | F

74   Cholesterol can be raised in anorexia nervosa.   T | F

75   Selective serotonin re-uptake inhibitors should be used with caution in patients over 80 years of age.   T | F

76   The suicide rate in prisoners is greatest immediately following reception into prison.   T | F

77   Tricyclic antidepressants cause fetal malformation if taken in pregnancy.   T | F

78   Piracetam is used as an adjunctive treatment for cortical myoclonus.   T | F

79   Being able to 'follow the evidence' is a requirement for fulfilment of the Pritchard criteria.   T | F

80   Trihexyphenidyl reduces the symptoms of Parkinsonism induced by antipsychotic drugs.   T | F

81   Grapefruit juice increases plasma concentrations of buspirone.   T | F

82   Aripiprazole was withdrawn in the 1960s because of an association with neutropenia and agranulocytosis.   T | F

83   Clozapine is reserved for treatment-resistant schizophrenia because of the cost implications.   T | F

84   The majority of psychiatric drugs are metabolised by cytochrome P450.   T | F

85   Schizophrenia is associated with cognitive impairment.   T | F

86   Agranulocytosis occurs in 2 per cent of patients in the first year of treatment with clozapine.   T | F

87   Treatment response rates to clozapine in patients having failed to respond to therapeutic trials of two antipsychotics (at least one being atypical) are 70–80 per cent.   T | F

88   Zotepine is one of only three atypical neuroleptics to lower the   T   F
     seizure threshold.

89   Breast-feeding whilst taking clozapine is acceptable if it is taken   T   F
     at the lowest therapeutic levels possible.

90   Aripiprazole causes clinically significant prolactin elevation when   T   F
     taken at therapeutic levels.

91   Risperidone is associated with increased risk of stroke in the   T   F
     elderly with dementia.

92   Attention-deficit hyperactivity disorder does not appear in ICD-10.   T   F

93   Studies show that attention deficit hyperactivity disorder in   T   F
     childhood is highly heritable.

94   Psychiatric comorbidity is almost 80 per cent in children with   T   F
     attention deficit hyperactivity disorder.   T   F

95   'Advance statements' are legally binding with regard to future   T   F
     healthcare choices should a serious illness develop which causes the
     individual to lose capacity or the ability to make their wishes known.

96   An elevated risk of suicide has been reported in association with   T   F
     migraine.

97   In people with psychosis, there is a marked excess of victimising   T   F
     experiences.

98   Individuals with schizotypal personality disorder use mental   T   F
     health services extensively.

99   Traditionally, an ECT-induced seizure should last 15 seconds on   T   F
     the EEG tracing to be therapeutic.

100 Up to 1 in 10 transpeople have a mental illness.   T   F

101 Motivational interviewing principles include confrontation rather   T   F
     than empathy.

102 Catatonic symptoms occur in autistic spectrum disorders.   T   F

103 Antidepressant monotherapy in bipolar depression is associated   T   F
     with worsening of symptoms.

104 Rapid cycling bipolar disorder describes six or more episodes of   T   F
     depression, mania, mixed state or hypomania in the preceding
     year.   T   F

105   Dialectical therapy was originally developed as a training manual   T | F
for treating dissocial personality disorder.

106   Selective serotonin re-uptake inhibitors are known to cause   T | F
arthralgia.

107   Tramadol enhances serotonergic and adrenergic pathways.   T | F

108   The risk of fatal agranulocytosis with clozapine treatment is 1 in   T | F
5000.

109   Vitamin E is a useful adjunctive treatment in Alzheimer's disease.   T | F

110   Ten per cent of bipolar patients develop the illness after the age   T | F
of 50.

111   Depression is likely to be a symptom of cognitive decline rather   T | F
than an independent risk factor.

112   The elderly have the second highest risk of completed suicide   T | F
when compared to other age groups.

113   The lifetime risk to an individual of developing a bipolar   T | F
disorder is approximately 5 per cent if they have no affected
relatives.

114   ECT treatment causes a rise in prolactin levels.   T | F

115   Depersonalisation disorder has an equal sex distribution.   T | F

116   Child-onset dysthymic disorder has a better outcome than major   T | F
depression.

117   Adults seem to be more sensitive than children to the effects of   T | F
trauma.

118   Lithium use in pregnancy is associated with fetal hypoglycaemia.   T | F

119   In neurosyphilis, the Argyll Robertson pupil is only seen in 15 per   T | F
cent of cases.

120   Symptoms of heroin withdrawal usually start 24 hours after the   T | F
last drug dose.

121   Lofexidine is a synthetic long-acting opiate used in opiate   T | F
withdrawal.

122   Selective serotonin re-uptake inhibitors are a cause of premature   T | F
ejaculation.

123   The stop–start technique is used for male premature   T | F
       ejaculation.

124   Abreaction is the emotional release or discharge after recalling a   T | F
       painful experience.

125   Pseudologia phantastica is a delusional belief that someone is   T | F
       deeply in love with you.

126   Amphetamine withdrawal is usually associated with mixed mood   T | F
       features.

127   Ecstasy is not associated with long-term brain damage.   T | F

128   Ultradian rapid cycling bipolar disorder relates to abrupt shifts in   T | F
       mood within a 72-hour period.

129   There is an association between ultra-rapid cycling bipolar   T | F
       disorder and 22q11.2 deletion syndrome.

130   Low body weight is a risk factor for antidepressant-induced   T | F
       hyponatraemia.

131   Social phobia has a lifetime prevalence of 1 per cent.   T | F

132   Memantine given together with an acetylcholinesterase inhibitor   T | F
       may benefit some patients with Alzheimer's dementia.

133   Psychological debriefing is the gold standard therapy in post-   T | F
       disaster counselling.

134   Central dopaminergic neurones reside mainly in the   T | F
       mesencephalon in three neuronal groups: the retrobulbar area, the
       substantia nigra and the ventral tegmental area.

135   Electra complex involves sexual feelings towards the mother   T | F
       whilst feeling hostility towards the father.

136   Methanol consumption is associated with 20 per cent mortality.   T | F

137   Patients with general paresis of the insane commonly present   T | F
       with grandiosity.

138   There is an over-representation of people with epilepsy in the   T | F
       prison population.

139   Unemployment is a risk factor for completed suicide in men.   T | F

140   The production of melatonin increases with age.   T | F

141   Chlorpromazine is a more potent antipsychotic than haloperidol.   T | F

142   Reboxetine is associated with weight gain.   T | F

143   Two per cent of patients with complex partial epilepsy have psychotic symptoms.   T | F

144   Catalepsy is a common feature of the narcolepsy syndrome.   T | F

145   Primary gain is achieved in conversion disorders.   T | F

146   Victim empathy is a widely used component of sex offender treatment.   T | F

147   Psilocybin is associated with hallucinogenic properties.   T | F

148   Self-mutilation occurs most commonly in the context of depression.   T | F

149   Seasonal affective disorder is associated with weight loss, reduced appetite and insomnia.   T | F

150   Light therapy for winter depression is most effective in the early morning.   T | F

151   Tryptophan depletion has been used to investigate depression.   T | F

152   ECT has been shown to be effective in childhood depression.   T | F

153   Cluttering is a disorder of speech usually occurring in childhood or adolescence.   T | F

154   Typically, in childhood selective mutism the child talks freely to his or her parents.   T | F

155   In sibling rivalry disorder, the older child can lose previously acquired skills.   T | F

156   ECT can precipitate epilepsy.   T | F

157   Childhood perfectionism is a risk factor for developing an eating disorder.   T | F

158   The brain-derived neurotrophic factor (BDNF) gene is thought to be a susceptibility gene for anorexia nervosa.   T | F

159   The theory of kindling relates to eating disorders.   T | F

160   The commonest symptom pattern in obsessive–compulsive disorder relates to pathological doubt.   T | F

161   Glutamate release is stimulated by nicotine.   T | F

162   Anger is recognised as the first phase in the psychological   T | F
       adjustment to dying.

163   Lycanthropy is the delusion of being a wolf.   T | F

164   Alexithymia is the inability to interpret the emotions of others.   T | F

165   Lithium causes thyroid enlargement in 5 per cent of patients.   T | F

EMIs

1

A	Prolactin	F	Parathormone
B	Somatostatin	G	Cortisol
C	Calcitonin	H	Thyrotropin releasing factor
D	Angiotensin	I	Substance P
E	Oxytocin	J	Neurotensin

Which of the above relates most closely to each of the statements below?

1   This is an adenohypophyseal hormone.

2   In the striatonigral pathway, this substance is thought to act as a neurotransmitter involved in pain perception.

3   Release is increased by nipple stimulation.

2

A	Redistribution	F	Volume of distribution
B	Zero-order kinetics	G	Therapeutic index
C	Half-life	H	Plasma index
D	Bioavailability	I	Acetylation
E	Steady state	J	Conjugation

Which term is being described in each of the following?

1   The fraction of administered drug reaching the systemic circulation without having been metabolised.

2   Is represented by the equation:
mass of drug in the body at a given time/plasma concentration of the drug at a given time.

3   A process in which the rate of drug elimination is constant.

3

A	Exhibitionism	F	Frotteurism
B	Fetishism	G	Paedophilia
C	Transsexualism	H	Voyeurism
D	Transvestic fetishism	I	Hypersexuality
E	Sexual masochism	J	Sexual sadism

Each description below relates to a diagnosis above.

1   Arousal by non-living objects, usually items of clothing or garments. These are often held, rubbed or smelt.

2   Sexual arousal by touching or rubbing against a non-consenting individual.

3   Sexual arousal from being hurt, humiliated, threatened or made to suffer in some way.

4

A	Pseudodementia	F	Lead intoxication
B	Vascular dementia	G	Lewy body dementia
C	Alzheimer's dementia	H	Huntington's dementia
D	Dementia in Parkinson's disease	I	Pick's disease
E	Alcohol-related dementia	J	Creutzfeldt–Jacob disease

In each of the following cases, which of the above is the most likely diagnosis?

1 A short history of rapidly progressive cognitive decline. Making little effort on cognitive testing, with an inconsistent performance.

2 Post-mortem examination has revealed intracytoplasmic neurofibrillary tangles, extracellular senile plaques, granulovacuolar degeneration and amyloid deposition in blood vessel walls.

3 A 54-year-old man has become increasingly disinhibited and lacking in judgement. His wife complains that he is increasingly apathetic and inappropriately jocular. He has reasonable memory.

4 A 73-year-old woman with a mini-mental score of 23/30. She has a resting tremor and showed marked sensitivity when given neuroleptic medication for visual hallucinations.

5 An EEG tracing shows asymmetry, localised slow waves and sparing of background activity.

5

A	Alzheimer's disease	F	Huntington's disease
B	Dementia with Lewy bodies	G	Creutzfeldt–Jacob disease
C	Pick's disease	H	Punch drunk syndrome
D	Vascular dementia	I	Paralysis agitans
E	Dementia in Parkinson's disease	J	Wilson's disease

Which of the above disorders relates to each of the descriptions below?

1 Knife blade gyri are characteristic.

2 An association with REM sleep behaviour disorder has only recently been described.

3 Perforation of the septum pallucidum is a macroscopic feature.

4 It is characterised by selective loss of discrete neuronal populations with progressive degeneration of efferent neurones of the neostriatum, with sparing of dopamine afferents.

6

A	Kraeplin 1894	F	Post 1966
B	Bleuler 1911	G	Grahame 1984
C	Mayer 1921	H	Murray-Parks 1988
D	Roth and Morrisey 1952	I	Kendell 1976
E	Fish 1960	J	Paykel 1971

Match each of the following to its rightful owner.

1   Separated four groups of depressives using cluster analysis.

2   Suggested that late paraphrenia is just schizophrenia in old age.

3   Described 'persistent persecutory states' – schizophrenic syndrome, schizophreniform syndrome and paranoid hallucinosis.

4   Dementia praecox, a disorder of emotion and volition.

5   Depression represents a continuum with varying degrees of melancholic and neurotic symptoms.

INDIVIDUAL STATEMENTS: Answers

1   **False.** This is the definition of a confounder.

2   **True.**

3   **True.**

4   **True.** For example 0.001, a more stringent criterion for rejecting the null hypothesis.

5   **True.** But the mathematics involves raw data, not proportions.

6   **False.** It is the other way round: clinically significant difference/standard deviation.

7   **False.** 1/absolute risk reduction.

8   **True.**

9   **True.**
    Sterne JA et al. *BMJ* 2001, **323**:101–5.

10 **True.** In other words, missing a true association.

11 **False.** A trial is double blind if neither subject nor investigator knows the treatment condition to which the subject has been randomised.

12 **True.**

13 **False.** This description is for predictive validity. Construct validity is the extent to which the construct that the measure looks to address is a real and coherent entity.

14 **True.**

15 **True.**

16 **False.** Five possible explanations exist: (1) reverse causality, (2) bias, (3) confounding, (4) chance, (5) causality.

17 **True.** In addition they are very time consuming, expensive and are generally unsuitable for the study of rare outcomes.

18   **True.** In general the larger the study, the greater its power – or the greater the power desired, the larger the study must be.

19   **True.**

20   **False.**

21   **True.**
Bradley H. *Am J Psychiatry* 1937, **94**:577–85.

22   **True.** Kanner's original paper describes 11 children seen over the course of 5 years who struck him as sharing fascinating peculiarities, including delayed echolalia, pronoun reversal, failure to use speech to communicate, an anxious desire to preserve sameness and repetitive behaviours.

23   **False.** It is present in 10 per cent of 5 year olds, 5 per cent of 10 year olds and 1 per cent of 15 year olds.

24   **False.** Mild learning disability 50–70, moderate 35–49, severe 20–34, and profound < 20.

25   **True.** Microcephaly, brachycephaly and flattened occiput are characteristic.

26   **True.** Other features are short fingers and clinodactyly (incurvature) of the fifth finger, which often only has two phalanges.

27   **True.** An additional technique is retraining the child with behavioural techniques, e.g. star charts. Washouts and manual evacuation are rarely needed.

28   **True.** Actus rea (guilty act), mens rea (guilty mind).

29   **False.** The McNaughten Rules (1843) require that 'at the time of committing the act, the party accused was labouring under such a defect of reason, from the disease of the mind, as not to know that what he was doing was wrong'.

30   **False.** Infanticide is the death of a child before 12 months of age caused by the mother through 'disturbed balance of her mind', having not fully recovered from the effect of lactation consequent upon the birth of the child (Section 1(1) Infanticide Act 1938).

31   **True.** The Pritchard Criteria: the defendant can understand the charge, instruct a lawyer, challenge a juror, plead to the charge and follow the evidence.

32   **False.** The figure is 10 per cent. For example conditions such as fragile X syndrome or tuberous sclerosis.

33   **False.** Rett's syndrome only occurs in girls. Features include regression and deceleration of head growth, handwashing stereotypies, episodic hyperventilation, unprovoked laughter and worsening mobility.

34   **True.** Patients who developed psychosis showed a reduction in grey matter in the left parahippocampal, fusiform, orbitofrontal and cerebellar cortices.
Pantelis C et al. *Lancet* 2003, **361**(9354):281–8.

35   **True.** A number of studies have shown this. One showed that adults with attention deficit hyperactivity disorder (ADHD) could be distinguished from healthy and non-ADHD individuals based on elevated theta activities, more total power and a higher theta:beta ratio. The researchers compared quantitative EEGs from 50 ADHD patients, 50 non-ADHD and 50 healthy participants.
Bresnehan S, Barry R. *Psychiatry Res* 2002, **112**(2):133–44.

36   **False.** This refers to the axis system employed by the *Diagnostic and Statistical Manual* (DSM).

37   **False.** Self-focused attention leads to increased state social anxiety, and task-focused attention reduces state social anxiety.

38   **True.** Inositol depletion has been implicated in the mechanism of action of these drugs and provides clues to the molecular basis of bipolar affective disorder.
Williams RS et al. *Nature* 2002, **417**:292–5.

39   **True.** Depression was associated with mortality for men but not for women in a study of 1947 men and women aged 70+. Depression may be an early sign of impending physical decline, or it may incur a physiological response that predisposes to cardiovascular disease or cancer.
Anstey KJ, Luszcz MA. *Psychosom Med* 2002, **64**:880–8.

40   **True.** Sildenafil citrate is Viagra.

41   **True.** Although at an early stage of research, tau levels were significantly higher in patients who progressed to probable Alzheimer's disease or progressive mild cognitive impairment.
Riemenschneider M et al. *Arch Neurol* 2002, **59**:1729–34.

42 **False.** In one study, 32 per cent had a diagnosis of depression and 42 per cent had depressive symptoms.
Chwastiak L et al. *Am J Psychiatry* 2002, 159:1862–8.

43 **True.**
Pallanti S et al. *J Clin Psychiatry* 2002, 63:1034–9.

44 **True.** Mothers who were under 35 years of age when their children with Down's syndrome were born were four to five times more likely to develop Alzheimer's disease than control mothers.
Schupf N et al. *Neurology* 2001, 57:979–84.

45 **True.** In a study of 422 people, those who originally had an unsteady gait were twice as likely to develop vascular dementia, while those with frontal gait had a more than fourfold increased risk. The highest risk was associated with hemiparetic gait: affected people were 13 times more likely to develop this type of dementia.
Verghese J et al. *N Engl J Med* 2002, 347:1761–8.

46 **True.**

47 **False.** Following the Gillick case, the courts have held that children with sufficient understanding and intelligence to enable them to understand fully what is involved in a proposed intervention will also have the capacity to consent to that intervention. Sometimes referred to as 'Gillick competent', a child under 16 may therefore have capacity to consent to some interventions but not to others.

48 **False.** It is usually in the 30s to 40s. Chromosome 4. Mutation of trinucleotide repeats (CAG). Normally there are 11–24 repeats, whereas in Huntington's disease there are 42–86 repeats.

49 **True.** Psychiatric manifestations include reactive affective disorders and personality disturbance including psychosis.

50 **False.** Lamotrigine is associated with potentially severe skin rashes which are minimised by gradual dose titration. It is used as adjunctive therapy, and may benefit rapid cyclers and bipolar depressed.

51 **True.** This was the first study showing changes in brain metabolism in chronically somatising women. It was a small study comparing only 10 cases with 17 controls.
Hakala M et al. *Psychol Med* 2002, 32:1379–85.

52   **True.** Also nausea, fatigue, restlessness, rash and photosensitivity. It is also called St John's wort. It can be purchased over the counter and is widely used as an antidepressant in Germany.

53   **False.** 5HT2a receptor expression was more abundant in the prefrontal cortex and hippocampus of adolescent suicide victims than in control subjects. However, this was a small study (15 suicide victims and 15 controls).
Ghanshyam N et al. *Am J Psychiatry* 2002, 159:419–23.

54   **True.** There is a significant association with adult schizophrenia (p = 0.042) and it is modestly associated with affective psychosis.
Leask SJ et al. *Br J Psychiatry* 2002, 181:387–92.

55   **False.** The association is with higher levels. Interleukin 6 is a pro-inflammatory cytokine with the capacity to induce a syndrome of 'sickness behaviour' sharing many of the features of depression, including anhedonia, fatigue, anorexia, reduced activity and altered sleep patterns.
Dominique L et al. *Am J Psychiatry* 2001, 158:1252–7.

56   **False.** A multicentre international study showed no millennium effect.
Sauer J et al. *Int J Soc Psychiatry* 2002, 48(2):122–5.

57   **False.** An open label study using paroxetine in depressed patients with Parkinson's disease did not appear to modify motor function.
Ceravolo R et al. *Neurology* 2000, 55:1216–18.

58   **False.** In a study of 101 patients with borderline personality disorder, 121 with post-traumatic stress disorder and 48 with both, having both did little to exacerbate existing pathology or dysfunction.
Zlotnick et al. *Am J Psychiatry* 2002, 159:1940–3.

59   **True.** However, donepezil was not shown to increase REM time or reduce slow-wave sleep.
Perlis ML et al. *Biol Psychiatry* 2002, 51:457–62.

60   **True.**
Twenge JM, Nolen-Hoeksema S. *J Abnorm Psychol* 2002, 111:578–88.

61   **True.** A small study showed a 46 per cent greater improvement compared to placebo. Selegiline is a monoamine-oxidase B inhibitor used in severe Parkinsonism in conjunction with levodopa to reduce 'end of dose' deterioration.
Bodkin JA, Amsterdam JD. *Am J Psychiatry* 2002, 159:1869–75.

62 **False.** Hydroxyzine is a sedating antihistamine. It has anti-anxiety properties and has been shown to be an alternative to benzodiazepines in some studies.
Llorca PM et al. *J Clin Psychiatry* 2002, 63:1020–7.

63 **True.**

64 **True.** These are equivalent to antidepressant effects.

65 **False.** It was derived from the daffodil.

66 **True.**

67 **False.** Supratentorial atrophy is seen.

68 **False.** There is an increased risk of mortality and stroke in elderly patients with dementia-related psychosis and/or behaviour disturbances. (Lily drug information leaflet 2nd March 2004.)

69 **True.**

70 **True.**

71 **True.**

72 **False.** It increases the QT interval.

73 **True.** An anti-arrhythmic, it may precipitate arrhythmias itself.

74 **True.**

75 **True.** They should also be used with caution in those with a history of upper gastrointestinal bleeding, those taking aspirin or another non-steroidal. Selective serotonin re-uptake inhibitors may interfere with clotting by inhibiting storage and the uptake of serotonin by platelets.
Drug Ther Bull 2004, 42(3):17–18.

76 **True.**
Shaw J et al. *Br J Psychiatry* 2004, 184:183–6.

77 **False.** Reported perinatal complications include withdrawal symptoms, irritability, eating and sleeping difficulties and convulsions.
McElhalton PR et al. *Reprod Toxicol* 1996, 10:285–94.

78 **True.**

79　**True.** The criteria relate to fitness to plead.

80　**True.** This is benzhexol.

81　**True.**
British National Formulary, 47th edn. Appendix 1: Interactions, anxiolytics and hypnotics. Buspirone used for anxiety (short term).

82　**False.** Clozapine was withdrawn in the 1960s following cases of neutropenia and agranulocytosis. Aripiprazole has recently gained a licence in the UK as an atypical neuroleptic. It is described as a dopamine stabiliser. Aripiprazole is an antipsychotic drug with high affinity for D(2) and D(3) receptors and the dopamine autoreceptor. It also has serotonin 5HT1a receptor partial agonist and 5HT2a receptor antagonist properties.

83　**False.** Because of adverse reactions.
Kilian JG et al. *Lancet* 1999, **354**:1641–5.

84　**True.**
Cozza K, Armstrong S. *The Cytochrome P450 System. Drug Interaction Principles for Medical Practice.* Washington, DC, American Psychiatric Publishing, 2001.

85　**True.** Schizophrenia is associated with *psychotic symptoms* (hallucinations, delusions, passivity experience), *disorganisation symptoms* (incongruous mood, abnormalities of speech and thought), *negative symptoms* (apathy, self-neglect, blunted mood, loss of motivation) and *cognitive impairment.*
Mueser KT, McGurk SR. *Lancet* 2004, **363**:2063–72.

86　**False.** It occurs in 0.7–0.8 per cent of patients.
Drug Ther Bull 1997, **35**:81–3.

87　**False.** Treatment response rates are 29–65 per cent.
Taylor D, Duncan-McConnell D. *J Psychopharmacol* 2000, **14**:409–18.

88　**False.** They all do.

89　**False.** It is absolutely contraindicated in breast-feeding due to the risk of agranulocytosis in the infant.

90　**False.** It is thought to have little effect on prolactin at therapeutic levels.

91　**True.** Also olanzapine.
Duff G. CEM/CMO/2004/1. Committee on Safety of Medicines, 2004.

92 **True.** Attention deficit hyperactivity disorder first appeared in DSM-III-R (American Psychiatric Association). ICD-10 includes hyperkinetic disorder.

93 **True.** It has a heritability of 0.8.
Taylor E et al. *Eur Child Adolesc Psychiatry* 1998, **7**:184–200.

94 **False.** It is around 60 per cent.
Green M et al. *Diagnosis and Treatment of ADHD in Children and Adolescents.* Technical Review, Number 3. Agency for Healthcare Policy and Research. AHCPR Publication Number 99-0050. Rockville, MD, American Medical Association, 1999.

95 **False.** 'Advance statements' are not legally binding, but should be honoured where possible. 'Advance directives' are legally binding if criteria for capacity and applicability are met at the time of completion.
Williams L, Rigby J. *Adv Psychiatr Treat* 2004, **10**(4):260–6.

96 **True.**
Breslau N et al. *Psychiatry Res* 1991, **37**:11–23.

97 **True.**
Bebbington P et al. *Br J Psychiatry* 2004, **185**:220–6.

98 **True.** Borderline and schizotypal personality disorders are associated with placing higher demands on mental healthcare services.
Bender DS et al. *Am J Psychiatry* 2001, **158**:295–302.

99 **False.** 25 seconds on EEG, 15 seconds peripherally.

100 **True.** One in ten have problems with mental illness, genital mutilation or suicide attempts. People who identify themselves with transsexualism are recognised as transpeople.
Wylie K. *BMJ* 2004, **329**:615–17.

101 **False.** The principles are:
empathy using reflective listening
develop discrepancy between deeply held values and existing behaviour
avoid resistance by empathising and understanding rather than confrontation
build the person's confidence that change is possible.
Miller WR, Rollnick S. *Motivational Interviewing: Preparing People for Change*, 2nd edn. Portland, OR, Book News Inc., 2002.

102 **True.** They can occur in autistic spectrum disorders, with a similar presentation to the catatonia associated with schizophrenia.
Wing L, Shah A. *Br J Psychiatry* 2000, **176**:357–62.

103 **True.** Antidepressant monotherapy can be associated with worsening of symptoms in bipolar disorder, including rapid cycling.

104 **False.** It describes four or more episodes.

105 **False.** Dialectical therapy was designed originally by M. Linehan for the treatment of borderline personality disorder. The therapist is supportive and directive.

106 **True.** Selective serotonin re-uptake inhibitors can cause arthralgia and myalgia as side effects.

107 **True.** It produces analgesia through this mechanism and also through its opioid effect. Psychiatric reactions have been reported with tramadol.

108 **True.** This is the main reason that clozapine-treated patients are monitored in the UK by the Clozaril Patient Monitoring Service. Other risks include pulmonary embolus, myocarditis and cardiomyopathy.

109 **False.** A recent Cochrane Review concludes that there is insufficient evidence to recommend vitamin E in Alzheimer's disease.
Tabet N et al. 2003, *Cochrane Database Syst Rev* 2000; (4):CD002854.

110 **True.**
Sajatovic M. *Int J Geriatr Psychiatry* 2002, **17**:865–73.

111 **True.** In a study of 500 elderly people in the Netherlands, impairment of attention/memory in old age preceded the development of depressive symptoms.
Vinkers DJ et al. *BMJ* 2004, **329**:881–3.

112 **False.** They have the highest risk, but lower absolute prevalence rates.
WHO 2002 <www.who.int/mental_health/prevention/suicide> (accessed 1st September 2004).

113 **False.** The lifetime risk is approximately 1 per cent; 5–10 per cent with an affected first-degree relative; 45–75 per cent for a monozygotic twin.

114 **True.** It causes a transient rise as with epileptic seizures. Other causes include medication (risperidone, amisulpride, zotepine). Medical causes include pituitary disease (prolactinomas), chronic renal failure, hypothyroidism, sarcoid. Physiological causes include pregnancy, breast-feeding and stress.

115 **True.** Depersonalisation disorder is characterised by prominent depersonalisation and often derealisation, without clinically notable memory or identity disturbances.
Simeon D. *CNS Drugs* 2004, **18**:343–54.

116 **False.** Childhood-onset dysthymic disorder has the worse outcome. It is characterised by a persistent and long-term depressed or irritable mood and is often associated with multiple problems and comorbidity.
Nobile M et al. *CNS Drugs* 2003, **17**:927–46.

117 **False.** Early life exposure may lead to multiple psychiatric disorders in adulthood.
Donnelly CL. *Child Adolesc Psychiatr Clin N Am* 2003, **12**(2):251–69.

118 **True.** Others reported include Ebstein's anomaly, poor respiratory effort, rhythm disturbances, nephrogenic diabetes insipidus, thyroid dysfunction, hypotonia and lethargy, hyperbilirubinaemia and large for gestational age infants.

119 **False.** It is seen in more than 50 per cent of cases. Depression is a common presenting feature. Treatment is with penicillin and steroids.

120 **False.** They usually start within 12 hours of the last dose, peaking within 72 hours, and can last up to 1 week.

121 **False.** Lofexidine is a centrally acting alpha-2 agonist used in reducing withdrawal symptoms.

122 **False.** They cause delayed ejaculation.

123 **True.** It is a behavioural treatment described by Semans.
Semans JH. *South Med J* 1956, **49**:3–8.

124 **True.**

125 **False.** The definition described relates to erotomania (de Cleramboult's syndrome). Pseudologia phantastica is a type of lying or story telling in which the individual believes the fantasies he tells others and acts on them. It is associated with Munchausen's syndrome.

126 **False.** Generally, intoxication is associated with manic features and withdrawal with depressive mood.

127 **False.** Ecstasy may cause long-term brain damage.
Mathins R. *NIDA Notes* 1996, **11**(5):7.

128 **False.** Ultradian rapid cycling (ultra ultra rapid cycling) relates to mood shifts of ≤ 24 hours' duration. Ultra rapid cycling relates to cycling of mood within the course of several days to weeks. Rapid cycling bipolar disorder demands four or more mood episodes in the previous 12 months. (Rapid-Cycling Specifier DSM-IV.)

129 **True.** Velocardio-facial syndrome/DiGeorge syndrome associated with catechol-O-methyltransferase (COMT) deletion.

130 **True.** Others include old age, poor renal function, medical comorbidity and other medications (e.g. carbamazepine). Seek medical advice if the sodium falls below 125 mmol/L and consider withdrawing the antidepressant.

131 **False.** The lifetime prevalence is 3–13 per cent.
Diagnostic and Statistical Manual. Washington, DC, American Psychiatric Association, 2000.

132 **True.** Whilst further studies are needed, recent research indicates that this might be true. A randomised controlled trial by Tarriot et al. showed significant improvement in physical and mental health in people with severe–moderate Alzheimer's disease when taking both memantine and donepezil.
Tarriot P et al. *JAMA* 2004, **291**:317–24.

133 **False.** Research has suggested both benefits and potential harmful effects from debriefing. It uses methods including abreaction, psychoeducation and cognitive restructuring in order to decrease arousal.

134 **True.** Retrobulbar area: regulation of hormone secretion from the pituitary. Substantia nigra: initiation and movement execution. Ventral tegmental area: limbic and limbic-connected areas involved in motivation, mood and organisation of thought.

135 **False.** The description is of the Oedipus complex. The Electra complex involves a female infant with sexual feelings towards the father whilst feeling hostility towards the mother.

136 **True.** Toxicity occurs through its metabolites, formaldehyde and formate. Treatment is with an ethanol infusion; 10 mL of methanol can cause blindness and 30 mL can be fatal.

137 **False.** They are more likely to present with psychosis, depression or dementia. General paresis of the insane affects around 50 per cent of people with neurosyphilis.

138 **True.** The reasons for this are not entirely clear; it is perhaps due to a degree of brain damage.

139 **True.**
Pritchard C. *Soc Psychiatry Psychiatr Epidemiol* 1988, 23(2):85–9.

140 **False.** Melatonin production decreases with age. Melatonin is produced during darkness by the pineal gland. It is used widely to minimise the effects of jet-lag. The therapeutic dose has not formally been determined.

141 **False.** Haloperidol has greater potency than chlorpromazine, as 5 mg of haloperidol has equivalent efficacy to 100 mg of chlorpromazine. Potency is the relative dose required to achieve a specific effect.

142 False. It may even attenuate olanzapine-associated weight gain.

143 **False.** It is estimated at 10–30 per cent of patients. Inter-ictal psychosis is commoner than ictal psychosis.

144 **False.** Cataplexy is often associated with narcolepsy. Catalepsy is a motor disturbance associated with schizophrenia. Patients often adopt an odd posture, which they maintain for extended periods of time.

145 **True.** Internal conflicts are kept outside of their awareness. They also often have secondary gain whereby they are excused from obligations and difficult situations in their lives.

146 **True.** Especially in the USA and UK. It involves enhancing empathy or compassion for victims.

147 **True.** 'Magic' *Psilocybe* mushrooms are associated with a hallucinogenic effect.

148 **False.** Self-mutilation occurs most commonly in the context of borderline personality disorder.
Langbehn DR, Pfehl B. *Ann Clin Psychiatry* 1993, 5:45–51.

149 **False.** Seasonal affective disorder is a subtype of affective disorder and is associated with increased weight, carbohydrate craving and hypersomnia.

150 **True.** The light box should be at least 10 000 Lux. Side effects of light therapy include headache, blurred vision, insomnia and over-activity.
Eagles J. *Adv Psychiatr Treat* 2004, 10(3):233–40.

151 **True.** Tryptophan depletion produces a marked reduction in plasma tryptophan and therefore brain serotonin synthesis and release.

152 **False.** It has not been thoroughly evaluated. It is rarely used.

153 **True.** ICD-10. Rapid rate of speech with poor fluency and reduced intelligibility. ICD-10 (F98.6).

154 **True.** Usually the children do not speak to classmates or teachers at school. ICD-10 (F94.0), elective mutism.

155 **True.** For example bowel or bladder control, as part of the emotional disturbance. ICD-10 (F93.3). In most cases the disturbance is mild and of little significance.

156 **True.** This is rare, but has been reported. Treat as for epilepsy.

157 **True.**
Anderluh MB et al. *Am J Psychiatry* 2003, **160**:242–7.

158 **True.** The associated protein regulates feeding behaviour in the hypothalamus.
Ribases M et al. *Hum Molec Genet* 2004, 13:1205–12.

159 **False.** It relates to mood disorders – to the dual effectiveness of anticonvulsants in both epilepsy and affective disorders. In epilepsy, kindling is thought to represent the repeated subthreshold neuronal stimulations in a brain area resulting in a seizure.

160 **False.** An obsession of doubt followed by a compulsion of checking is the second commonest. The commonest is the obsession of contamination.

161 **True.** It is the major excitatory amino acid.

162 **False.** According to Kübler-Ross, the phases are (1) denial and isolation, (2) anger, (3) bargaining, (4) depression, (5) acceptance.

163 **True.** Defined by Karl Jaspers, who contributed extensively to phenomenological psychopathology.
Jaspers K. *General Psychopathology.* (Translated from the 7th German edition by Hoenig J, Hamilton MW.) Manchester, Manchester University Press, 1963.

164 **False.** It is the inability to recognise and describe one's own feelings or emotions.
Nemiah JC, Sifneos PE. *Psychother Psychosom* 1970, **18**(1):154–60.

165 **True.** It is reversible on withdrawal of lithium.

EMIs: Answers

1

1   **A**   Anterior pituitary hormones are ACTH, FSH, LH, melanocyte stimulating hormone, prolactin, GH and TSH. Thyrotropin releasing factor is an anterior pituitary releasing factor, so not strictly speaking a hormone.

2   **I**   Substance P has been hypothesised to be involved in mood disorders and dementia.

3   **E**   Oxytocin and vasopressin have been proposed as being important in mood regulation.

2

1   **D**

2   **F**

3   **B**

3

1   **B**   The paraphilias are predominantly male conditions.

2   **F**

3   **E**

4

1   A   Typically 'I don't know' answers. Examine for depressive features.

2   C   Typical Alzheimer's pathology.

3   I   It is important to examine frontal lobes and frontal release signs.

4   G   Often diagnosed as dementia in Parkinson's disease. There is overlap between the two diagnoses.

5   B

5

1   C   Also there is a relative asymmetrical atrophy of anterior temporal and frontal lobes.

2   B   REM sleep behaviour disorder: motor movements occurring in REM sleep.

3   H   As is cerebral atrophy, ventricular enlargement and thinning of the corpus callosum.

4   F

6

1   J   Paykel's classification: (1) psychotic depressives, (2) anxious depressives, (3) hostile depressives, (4) younger depressives with personality disorders.

2   G

3   F

4   A

5   I

Note. Roth and Morrisey separated groups of depressives using multiple regression analysis, and argued that the melancholic and neurotic groups were distinct.

INDEX

Note: References are given in the form of the paper number in brackets followed by the question number. Rounded brackets indicate Individual Statements and square brackets the EMIs. 'vs' indicates the comparison of two items, most often the differential diagnosis of two conditions.